CONTROLLING VOLATILE ORGANIC COMPOUND EMISSIONS FROM INDUSTRIAL WASTEWATER

by

Jeffrey Elliott, Sheryl Watkins

Radian Corporation
Research Triangle Park, North Carolina

NOYES DATA CORPORATION
Park Ridge, New Jersey, U.S.A.

Library of Congress Catalog Card Number: 90-44910
ISBN: 0-8155-1261-9
ISSN: 0090-516X
Printed in the United States

Published in the United States of America by
Noyes Data Corporation
Mill Road, Park Ridge, New Jersey 07656

10 9 8 7 6 5 4 3 2 1

Library of Congress Cataloging-in-Publication Data

Elliott, Jeffrey.
Controlling volatile organic compound emissions from industrial wastewater / by Jeffrey Elliott, Sheryl Watkins.
p. cm. -- (Pollution technology review, ISSN 0090-516X ; no. 197)
Includes bibliographical references and index.
ISBN 0-8155-1261-9 :
1. Sewage--Purification--Organic compounds removal. 2. Factory and trade waste--Purification. I. Watkins, Sheryl. II. Title.
III. Series.
TD758.5.075E45 1990
628.1'683--dc20 90-44910
CIP

Foreword

This book presents a description of sources of organic containing wastewater, volatile organic compound (VOC) emission estimation procedures for treatment and collection system units, and available VOC emission control strategies. In addition, secondary impacts and the control costs associated with steam stripping are presented.

Technical information is provided to industry and governmental organizations for (1) estimating emissions of VOCs from the collection and treatment of industrial wastewaters; and (2) Best Available Control Technology (BACT), and Lowest Achievable Emission Rate (LAER) determinations for controlling emissions of VOCs from industrial wastewaters.

Although this book applies to four industries in particular: organic chemicals, plastics and synthetic fibers; pesticides; pharmaceuticals; and hazardous waste, treatment, storage and disposal—the technical information is applicable to a broad range of other industries.

Facilities within these industries have the potential to generate wastewaters containing high concentrations of organic compounds. These wastewaters typically pass through a series of collection and primary treatment units, before treatment is applied to reduce the concentration of organic compounds prior to discharge. Many of these collection and treatment units are open to the atmosphere and allow organic-containing wastewaters to contact ambient air. Atmospheric exposure of these organic-containing wastewaters results in significant volatilization of VOCs from the wastewater.

These emissions can be reduced by applying one of three control strategies. The most effective strategy is to apply waste minimization techniques to reduce the organic concentration of the wastewaters, or to produce a more manageable wastewater stream through waste segregation or recycling. Even with waste minimization, some waste streams will be generated. Emissions from these wastewater streams can be reduced by applying treatment at the point of generation. Numerous controls are suitable in specific cases, but the most universally applicable treatment technology for controlling emissions from wastewater generated by these industries is steam stripping. A third control strategy that may be appropriate for some situations is to enclose the wastewater collection system and cover all treatment units, up through removal or destruction of the organic compounds.

The information in the book is from *Industrial Wastewater Volatile Organic Compound Emissions—Background Information for BACT/LAER Determinations,* prepared by Jeffrey Elliott and Sheryl Watkins of Radian Corporation for the U.S. Environmental Protection Agency, January 1990.

The table of contents is organized in such a way as to serve as a subject index and provides easy access to the information contained in the book.

Advanced composition and production methods developed by Noyes Data Corporation are employed to bring this durably bound book to you in a minimum of time. Special techniques are used to close the gap between "manuscript" and "completed book." In order to keep the price of the book to a reasonable level, it has been partially reproduced by photo-offset directly from the original report and the cost saving passed on to the reader. Due to this method of publishing, certain portions of the book may be less legible than desired.

Acknowledgment

The Industrial Wastewater Volatile Organic Compound Emissions–Background Information for BACT/LAER Determinations Revised Draft was prepared for EPA's Control Technology Center (CTC) by J. Elliott and S. Watkins of Radian Corporation. The work assignment manager was Penny Lassiter of the EPA's Office of Air Quality Planning and Standards (OAQPS). Also participating on the project team were Bob Blaszczak, OAQPS and Chuck Darvin, Air and Energy Engineering Research Laboratory (AEERL).

NOTICE

The materials in this book were prepared as accounts of work sponsored by the U.S. Environmental Protection Agency. On this basis the Publisher assumes no responsibility nor liability for errors or any consequences arising from the use of the information contained herein. Mention of trade names or commercial products does not constitute endorsement or recommendation for use by the Agency or the Publisher.

The book is intended for informational purposes only. The reader is cautioned to obtain expert advice before implementation of any procedures described, since wastewaters to be controlled contain potentially hazardous volatile organic compounds. Final determination of the suitability of any information or procedure for use by any user, and the manner of that use, is the sole responsibility of the user.

All information pertaining to law and regulations is provided for background only. The reader must contact the appropriate legal sources and regulatory authorities for up-to-date regulatory requirements, and their interpretation and implementation.

Contents and Subject Index

1. Introduction

Under the prevention of significant deterioration (PSD) provisions of Part C of the Clean Air Act (CAA), a new major stationary source or a major modification shall apply best available control technology (BACT) for each pollutant regulated under the CAA that it would have the potential to emit in significant amounts. Similarly, under the nonattainment new source review (NSR) provisions of Part D of the CAA, a major new or modified source in a nonattainment area shall apply controls to attain the lowest achievable emission rate (LAER). A new major stationary source refers to any source within these source categories which emits, or has the potential to emit, 100 tons per year of VOC. A major modification refers to a physical change in or a change in the method of operation of a stationary source which results in a net increase in potential emissions of 40 tons per year of VOC.

The purpose of this document is to provide technical information to industry and States on 1) estimating emissions of volatile organic compounds (VOC) from the collection and treatment of industrial wastewaters, and 2) BACT and LAER determinations for controlling emissions of VOC from industrial wastewaters. This document applies to four industries: the Organic Chemicals, Plastics, and Synthetic Fibers (OCPSF) Industry; the Pesticide Industry, the Pharmaceutical Industry; and the Hazardous Waste Treatment Storage, and Disposal Facilities Industry (TSDF). However, this list could be expanded to include additional industries as information becomes available.

Based on information collected by the Agency, facilities within each of these industries have the potential to generate wastewaters containing high concentrations of organic compounds. These wastewaters typically pass through a series of collection and primary treatment units before treatment is applied to reduce the concentration of organic compounds prior to discharge. Many of these collection and treatment units are open to the atmosphere and allow organic-containing wastewaters to contact ambient air. Atmospheric exposure of these organic-containing wastewaters results in significant volatilization of VOC from the wastewater.

These emissions can be reduced by applying one of three control strategies. The most effective strategy is to apply waste minimization techniques to reduce the organic concentration of the wastewaters, or to produce a more manageable wastewater stream through waste segregation or recycling. Even with waste minimization, some waste streams will be generated. Emissions from these wastewater streams can be reduced by applying treatment at the point of generation. Numerous controls are suitable in specific cases, but the most universally applicable treatment technology for controlling emissions from wastewater generated by these industries is steam stripping. A third control strategy that may be appropriate for some situations is to enclose the wastewater collection system and cover all treatment units up though removal or destruction of the organic compounds.

The organization of this document is as follows. A description of the industries covered by this document is presented in Chapter 2. The sources of organic containing wastewater, and the sources of VOC air emissions are identified in Chapter 3. Chapter 4 presents available VOC emission control strategies. The secondary impacts and control costs associated with the recommended control strategy, steam stripping, are presented in Chapters 5 and 6, respectively.

2. Industry Descriptions

This Control Technology Center (CTC) document is intended to apply to industrial wastewater generated by new and modified sources within the following industries:

- The Organic Chemicals, Plastics, and Synthetic Fibers Manufacturing Industry, (OCPSF);
- The Pesticides Manufacturing Industry;
- The Pharmaceuticals Manufacturing Industry; and
- The Hazardous Waste Treatment, Storage, and Disposal Facilities Industry, (TSDF).

In addition, although not covered by this document, the EPA is in the process of gathering data on the Pulp, Paper and Paperboard and Builders Paper and Board Mill Industry (Pulp and Paper Industry). Information is presented in this chapter on wastewater streams generated by these five industries.

The industry descriptions and wastewater characteristics presented in this chapter reflect data collected by the EPA on VOC emissions from industrial wastewater, and work done by the EPA either to develop effluent guidelines or to evaluate the need to develop effluent guidelines.[1-7] The four industries listed above are included together in this document because based on the available data, their wastewaters are similar in characteristic and would have similar control requirements. The wastewater characteristics such as solids content, organic content, volatility and solubility of organic constituents, and wastewater volumes are all similar for pharmaceutical, pesticide, and OCPSF industries as evidenced by the use of steam strippers at facilities within all of these industries. Since a significant portion of the waste handled by the TSDF facilities is from one of these three industries, these wastewaters are also similar in characteristics. The similarities of these wastewaters are further supported by statements made in the Federal Register notice promulgating regulations for the pesticide industry.[8] Processes in these industries (pharmaceutical and OCPSF) are similar to those in the pesticides industry and the proposed effluent limitations guidelines and standards for the OCPSF, pharmaceuticals, and pesticides categories are based on the same treatment technologies. This list of these four industries could expand to include the pulp and paper industry as well as other industries as information becomes available.

Based on information gathered by EPA in developing effluent guideline limitations, there were approximately 3,500 facilities in 1982 in the four industries included in this document. These facilities produced approximately 700 million gallons per day (MGD) of wastewater. Table 2-1 presents estimates of the number of facilities and the quantities of wastewater generated by each industry. Based on available flow and concentration data, the quantity of organic compounds in wastewater generated by each of the four industries are considered significant. In addition, based on the information available on wastewater characteristics within these industries, similar controls are applicable and similar control requirements are warranted for new and modified sources within each industry.

Data collected in developing effluent guideline limitations were obtained from responses to questionnaires sent under the authority of Section 308 of the Clean Water Act (CWA) and field sampling and analysis. These data are typically restricted to 126 pollutants called priority pollutants. Of these, only 27 are VOC (as listed in Table 2-2). As a result, these data represent a subset of VOC; those organics that are also priority pollutants. Priority pollutant data may not, therefore, provide an accurate estimate of the total organic concentration in wastewater.

The EPA also collected data under the authority of Section 114 of the Clean Air Act to evaluate emissions of VOC and potentially hazardous air pollutants from the OCPSF industry. Testing was not required as part of the Section 114 request, and the data provided represent a combination of actual sampling data and engineering estimates. Information was also collected through site visits conducted by EPA to facilities within each of the industries included in this chapter. During this study, the correlation between VOC and priority pollutants was found to be highly variable with VOC content, typically one to six orders of magnitude greater than organic priority pollutant content for these industries.

The following sections discuss each of the four industries included in this document, and the pulp and paper industry, in terms of approximate number of facilities and the number of processes or products. Wastewater characteristics such as the sources and quantities of wastewater generated, and the methods of discharge are also provided.

TABLE 2-1. WASTEWATER GENERATION BY INDUSTRY

Industry	Total Number Of Facilities (1982)	Daily Wastewater Generation (MGD)
OCPSF	1,000	500
Pesticides Manufacturing	119	<100
Pharmaceuticals Manufacturing	466	93
TSDF	1,909	16[a]
TOTAL	3,500	700

[a]This only represents wastewater generated by the TSDF as landfill leachate. Actual quantities of organic-containing wastewater handled is much higher.

TABLE 2-2. PRIORITY ORGANIC COMPOUNDS THAT ARE VOC

Chemical	CAS Number
Acrolein	107-02-8
Acrylonitrile	107-13-1
Benzene	71-43-2
Bromodichloromethane	75-27-4
Bromoform	75-25-2
Bromomethane	74-83-9
Carbon tetrachloride	56-23-5
Chlorobenzene	108-90-7
Chloroethane	75-00-3
2-Chloroethylvinyl ether	110-75-8
Chloroform	67-66-3
Chloromethane	74-87-3
Dibromochloromethane	124-48-1
1,1-dichloroethane	75-34-3
1,2-dichloroethane	107-06-2
1,1-dichloroethene	75-35-4
trans-1,2-dichloroethene	156-60-5
1,2-dichloropropane	78-87-5
cis-1,3-dichloropropane	10061-01-5
1,3-dichloropropane	10061-02-6
Ethylbenzene	100-41-4
1,1,2,2-tetrachloroethane	79-34-5
Tetrachloroethane	127-18-4
Toluene	108-88-3
1,1,2-trichloroethane	79-00-5
Trichloroethene	79-01-6
Vinyl chloride	75-01-4

Adapted from References 6.

2.1 ORGANIC CHEMICALS, PLASTICS, AND SYNTHETIC FIBERS MANUFACTURING

Approximately 1,000 facilities are included in the OCPSF industry, defined as all facilities falling under the following standard industrial classification (SIC) codes:

- 2821 Plastics Materials, Synthetic Resins, and Nonvulcanizable Elastomers;
- 2823 Cellulosic Manmade Fibers;
- 2824 Manmade Organic Fibers, except Cellulosic;
- 2865 Cyclic Organic Crudes and Intermediates, and Organic Dyes and Pigments; and
- 2869 Industrial Organic Chemicals, Not Elsewhere Classified.

The OCPSF industry includes a diversity of chemical processes producing a large number of chemical products. Some facilities within these industrial categories produce large volumes of a single product continuously while other facilities may produce various specialty products in short campaigns. However, despite the diversity of this industry, EPA has determined that 98 percent of all products manufactured are produced by one of 41 major generic processes. These processes are listed in Table 2-3. The OCPSF industry generates about 530 MGD of wastewater. About 32 percent of the OCPSF facilities are direct dischargers (i.e., wastewater is treated on-site and discharged directly to a water body); 42 percent are indirect dischargers (wastewater is discharged to a publicly owned treatment works (POTW)), and 26 percent are zero dischargers (no wastewater discharged from the facility). Estimates for average daily process wastewater flow per plant are 1.22 MGD for the direct dischargers and 0.24 MGD for the indirect dischargers.[9] The majority of this volume is from cooling water use. Most of the wastewater collection systems at facilities in the OCPSF industry are underground sewers. Very few wastewater streams are transported in overhead pipe. In addition, in some facilities, vigorous aeration of the wastewater prior to biological treatment is used to improve the biological activity. Based on OAQPS visits to several facilities in the OCPSF industry, significant potential exists in this industry for emissions of VOC from wastewater.

TABLE 2-3. GENERIC CHEMICAL PROCESSES

1.	Acid Cleavage	22.	Extractive Distillation
2.	Alkoxylation	23.	Fiber Production
3.	Alkylation	24.	Halogenation
4.	Amination	25.	Hydration
5.	Ammonolysis	26.	Hydroacetylation
6.	Ammoxidation	27.	Hydrodealkylation
7.	Carbonylation	28.	Hydrogenation
8.	Chlorohydrination	29.	Hydrohalogenation
9.	Condensation	30.	Hydrolysis
10.	Cracking	31.	Isomerization
11.	Crystallization/Distillation	32.	Neutralization
12.	Cyanation/Hydrocyanation	33.	Nitration
13.	Dehydration	34.	Oxidation
14.	Dehydrogenation	35.	Oximation
15.	Dehydrohalogenation	36.	Oxyhalogenation
16.	Distillation	37.	Peroxidation
17.	Electrohydrodimerization	38.	Phosgenation
18.	Epoxidation	39.	Polymerization
19.	Esterification	40.	Pyrolysis
20.	Etherification	41.	Sulfonation
21.	Extraction		

Taken from Reference 1.

Concentrations of organic pollutants are highly variable in process wastewater generated by OCPSF industry facilities. Table 2-4 presents some of the concentration data provided by facilities in response to Clean Air Act Section 114 requests. Table 2-4 lists the pollutants identified in more than three different wastewater streams and the minimum and maximum concentration reported for each pollutant. The largest range in concentration was reported for ethanol. The minimum and maximum concentrations reported for this pollutant were 199 milligrams per liter (mg/l) and 443,213 mg/l, respectively. Wastewaters generated in the OCPSF industry may contain moderate levels of oil and grease or suspended solids but usually not in levels which would preclude steam stripping. Steam stripping is an effective technology in use at many facilities in the OCPSF industry for removing organics from wastewater.

Table 2-4 is based on data gathered for this project and is provided for illustration only. The data is not necessarily all inclusive with regard to the compounds shown or the concentration ranges presented. The data simply illustrate that these wastewaters contain a variety of compounds with wide variations in concentration. Although concentrations for various pollutants are highly variable, the data indicate that a small number of streams contribute the majority of the organic compounds in OCPSF wastewater. Organic quantities were computed for each process wastewater stream where data were available from facility responses to the Section 114 information requests. In addition, a total organic quantity representing all the reported streams was determined by summing the organic quantities computed for each individual wastewater stream. Based on these data collected from facilities in the OCPSF industry, approximately 20 percent of the individual wastewater streams were found to account for 65 percent of the total organic quantity for all the reported streams.[10]

2.2 PESTICIDES MANUFACTURING

The Pesticides Manufacturing Industry provides a wide range of chemicals used to control crop-destroying insects and undesirable vegetation. This document covers the segment of the pesticide industry that manufactures pesticide chemical active ingredients. One hundred and nineteen such plants were identified in development of the 1985 effluent standards (50 FR 40674, October 4, 1985).[8]

TABLE 2-4. POLLUTANT CONCENTRATION DATA PRESENTED IN SECTION 114 RESPONSES

Pollutant	Concentration (mg/l)	
	Minimum	Maximum
1,3-Butadiene	1	2,986
Acetaldehyde	20	2,180
Acetone	14	1,220
Acrylonitrile	50	15,000
Butane	700	1,000
Butene	700	1,000
Perchloroethylene	24	100
Ethylene dichloride	5.44	10,110
Trichloroethane	50	100
Dichloroethane	273	10,110
Ethylene Oxide	2	1,252
Chloroethane	2	300
Benzene	0.44	2,800
Carbon tetrachloride	1	29,592
Methylene chloride	0.01	83
Chloroform	1	400
Chlorobenzene	0.1	50
Ethanol	199	443,213
Ethylbenzene	0.4	25
Formaldehyde	17,485	19,487
Gasoline (C5's thru C10's)	1,000	50,000
Hexane	90	10,000
Isobutane	0.1	0.2
Methylisobutyl Ketone	8,888	12,072
Naphtha	398	14,423

(CONTINUED)

TABLE 2-4. (CONTINUED)

Pollutant	Concentration (mg/l) Minimum	Maximum
Naphthalene	0.4	21
Phenol	6.2	125
Propylene Oxide	16.4	2,006
Styrene	5	25,524
Toluene	1	25,000
Total Organic Compounds	0.05	1,000,000
Triethylamine	1,622	100,000

These plants produce pesticide products covered under the following SIC codes:

2819	Industrial Inorganic Chemicals, Not Elsewhere Classified
2869	Industrial Organic Chemicals, Not Elsewhere Classified
2879	Pesticides and Agricultural Chemicals, Not Elsewhere Classified

(Nonpesticide products under these SIC codes are excluded.)

The volume of wastewater discharged by facilities in this industry ranges from less than 10,000 gallons per day to 1 MGD, with over half the facilities in the industry generating less than 10,000 gallons per day. Discharge methods vary from plant to plant and one method or a combination of methods may be used. Forty-five facilities directly discharge wastewater, 38 are indirect dischargers, and 18 are zero dischargers.

A variety of organic compounds have been detected in pesticides industry wastewater streams. These include: phenols, aromatics, halomethanes, clorinated ethanes, nitrosamines, dienes, cyanides, and pesticide compounds. Sampling data generated during effluent guidelines development on organic concentrations for the industry are limited to organic priority pollutant concentration data. High concentrations of halomethanes and chlorinated ethanes were detected in the pesticide plant wastewaters. The organic compounds detected in the wastewaters are used as solvents and raw materials or occur as impurities or by-products. The sources and characteristics of wastewaters generated by pesticide manufacturing facilities is expected to be similar to those in the OCPSF industry. Although the wastewater flow rates from many of the facilities is lower, the organic content and other wastewater characteristics, such as total suspended solids concentrations, appear to be similar to those in the OCPSF industry. The pesticide development document provides detailed information on ten steam strippers in use at Pesticide Industry plants.[3]

2.3 PHARMACEUTICALS MANUFACTURING

The Pharmaceutical Manufacturing Industry includes facilities which manufacture, extract, process, purify, and package chemical materials to be used as human and animal medications. Four hundred and sixty-six facilities were identified by EPA as Pharmaceutical Manufacturers. This industry as defined in the Federal Register (48 FR 49809, October 27, 1983)[11] includes facilities in the SIC codes:

2831 Which was split in 1987 into:

- 2835 Diagnostic Substance;
- 2836 Biological Products Except Diagnostic Substances;
- 2833 Medicinal Chemicals and Botanical Products;
- 2834 Pharmaceutical Preparations; and
- 2844 Perfumes, Cosmetics and Other Toilet Preparations which function as skin treatment.

Other facilities covered by this document are:

- The manufacture of products considered pharmaceutically active by the Food and Drug Administration;
- The manufacture of non-pharmaceutical products made at pharmaceutical manufacturing facilities that generate similar wastewater to those from pharmaceuticals production;
- The manufacture of products "which have non-pharmaceutical uses" but that were "primarily intended for use as a pharmaceutical"; and
- Pharmaceutical research.

Pharmaceutical production operations may be batch, semi-continuous, or continuous. However, the most common method of operation is batch. Manufacturing in the industry can be characterized by four processes. These are fermentation; extraction; synthesis; and mixing, compounding, or formulating.

Fermentation is usually a large scale batch process and involves fermentation, or controlled growth of specific microorganisms in a reactor vessel to produce a desired product. The desired product is then recovered from the fermentation broth using solvent extraction, adsorption, precipitation and filtration, or ion exchange. Waste streams generated from fermentation processes include discharges from reactor cleanings and sterilizations, off-gas scrubber effluents, and occasional off-specification batches. Solvents used in extracting the product from the broth in the recovery process may be discharged into the sewers in the waste streams as well.

Extraction refers to the extraction and recovery of a small volume of desired product from naturally occurring sources such as plant roots and leaves, animal glands, and parasitic fungi. Extraction operations are usually either batch or semi-continuous. Wastewater discharges from extraction processes include spent raw materials, solvents used in extractions, and spills and equipment wash waters.

Chemical synthesis, either through batch or continuous processes (usually batch), is the most common method of preparing pharmaceuticals. Synthesis of pharmaceuticals involves reaction of the appropriate raw materials and recovery of the desired product. Effluents from synthesis operations are highly variable as are the processes where they are generated. Process solutions, vessel wash wasters, filtrates, concentrates, spent solvents, and scrubber effluents are all sources of wastewater. Pump seal water, spills, and cleaning wash waters are additional sources. Any of these sources may contain significant concentrations of organics.

Mixing, compounding, and formulating operations involve preparation of the active ingredients into a dosage form for consumer use. The primary sources of wastewater from these processes are from equipment washings, scrubber effluents, and spills.

Although wastewater streams from all four processes have the potential to contain high organic loadings, fermentation and synthesis operations usually generate larger volumes of wastewater and the wastewaters generated usually contain higher organic loadings. Based on data gathering efforts by EPA, the pharmaceutical manufacturing industry discharges significant quantities of organic compounds in their raw wastewaters.

Discharge practices across the industry vary; 59 percent are indirect dischargers, 29 percent are zero dischargers, and the remainder are direct dischargers. Flow data from Clean Water Act Section 308 questionnaires are limited to the 70 percent of direct and indirect dischargers who provided responses. Over half the total 80 MGD generated by these respondents resulted from direct discharges (25 facilities contributed 45 MGD). To better determine the total industry wastewater generation, EPA estimated the contribution from the non-respondents at 13 MGD. The total wastewater flow, therefore, is approximately 93 MGD.

A study by the Pharmaceutical Manufacturers' Association, which focused on 26 member companies identified a total of 46 volatile organics used by the industry.[12] These companies represent 53 percent of the domestic sales of prescription drugs. Industry use of organic compounds is primarily as raw materials or solvents. An estimated 84 percent of the organic compounds are recycled and 16 percent are waste organics. Approximately 2.7 percent of the waste organics are discharged to the sewer.

Because much of the industry uses batch operations, this industry has more variability in its wastewater flows and organic content. However, in many cases these wastewaters contain large quantities of organic compounds. Information gathered on the characteristics and volumes of wastewater generated in this industry support the inclusion of this industry in this document. This is supported by the use of steam strippers by at least eight pharmaceutical manufacturers.[4]

2.4 HAZARDOUS WASTE TREATMENT, STORAGE, AND DISPOSAL FACILITIES (TSDF)[7]

In 1986, EPA conducted a study to gather data on wastewater produced by this source category as part of Clean Water Act effluent guidelines development work. The Domestic Sewage Study, performed by the EPA in response to Section 3018(a) of the Resource Conservation and Recovery Act (RCRA), identified TSDF (referred to as Hazardous Waste Treaters (HWT) in that study) as significant contributors of hazardous wastes to POTW. The EPA has placed very high priority on development of pretreatment standards for Centralized Waste Treaters to control toxic and hazardous pollutants.

EPA has divided the TSDF industry into three categories for effluent guideline purposes:

1. Landfills with leachate collection, including commercial (off-site) and industrial (on-site) hazardous waste (Subtitle C of RCRA) and municipal nonhazardous waste (Subtitle D of RCRA) landfills.

2. Hazardous waste incinerators with wet scrubbers (commercial and industrial).

3. Centralized waste treaters, including commercial, industrial and federal (Subtitle C of RCRA) TSDF with and without categorical effluent regulations.

EPA has identified 1,304 out of 1,909 facilities that would be subject to effluent guideline regulations if EPA develops any in the future. They break down as follows:

Facility Type	Direct Discharge	Indirect Discharge	Other*
Landfill Leachate	173	355	383
Incinerator Scrubber	137	27	109
Centralized Waste Treaters	87	515	123
TOTAL	397	907	605

----

*"Other" includes off-site disposal at a commercial aqueous waste treatment facility, deep well injection, and other methods.

Landfill leachates contain high concentrations of toxic organic compounds and metals, and conventional and nonconventional pollutants. Many organic compounds are in the range of 1 - 10 mg/l, a few at greater than 100 mg/l. Total mass in raw wastewater discharges of nonpriority organic compounds range from 1.8 to 4.7 times greater than priority organic compounds. This industry produces about 16 MGD of landfill leachate.

RCRA Subtitle D Surface Impoundments.[13] Subtitle D surface impoundments are impoundments that accept wastes as defined under Subtitle D of RCRA. Subtitle D wastes are all solid wastes regulated under the RCRA that are not subject to hazardous waste regulations under Subtitle C. These wastes are defined in 40 CFR Part 257. Specifically, this document applies to process wastewater produced by generators; small quantity generators; POTW; and TSDF that is RCRA Subtitle D waste as defined in 40 CFR 257.

These Subtitle D surface impoundments might be used for evaporation, polishing, storage prior to further treatment or disposal, equalization, leachate collection, emergency surge basin, etc. They could be quiescent or mechanically agitated.

A Subtitle D census conducted in 1986 identified 191,822 active surface impoundments located at 108,383 facilities. The results show that 16,232 Subtitle D surface impoundments accept industrial wastes.

The total nonhazardous waste generation was estimated to be roughly 390,000,000 metric tons per year, with 93 percent of this provided by seven industries: industrial organic chemicals; primary iron and steel; fertilizers and other agricultural chemicals; electric power generation; plastic and resin manufacture; industrial inorganic chemicals; and clay, glass, and concrete products. Table 2-5 shows the quantity of industrial waste managed in Subtitle D surface impoundments. Table 2-6 shows the number of Subtitle D surface impoundments accepting industrial waste by acreage category. The State Subtitle D Consensus estimated that 17,159 Subtitle D surface impoundments receive industrial wastes from small quantity generators. A breakdown of total small quantity generator waste by industry is shown in Table 2-7.

Incinerator wet scrubber liquors contain high concentration of toxic metals but very few organics at relatively low concentrations. Approximately 15 MGD of incinerator wet scrubber liquors are produced.

Centralized waste treatment facilities typically have high concentrations of toxic metals and organics. Numerous metals are found from 25 to 1,300 mg/l. Numerous priority organics are found greater than 1 mg/l and some greater than 10 mg/l. Total mass in raw wastewaters of nonpriority organics are approximated to be seven times greater than priority organics. Centralized waste treaters produce approximately 27 MGD.

2.5 PULP, PAPER AND PAPERBOARD AND BUILDERS' PAPER AND BOARD MILLS INDUSTRY

The industry totaled 695 operating facilities at the time of the EPA development work on the effluent limitations (final rule: 47 FR 52006, November 18, 1982[14]). A more recent estimate prepared by EPA in 1989 indicates that there are now 603 facilities in this industry. Table 2-8 shows a breakdown of the facilities by subcategory as estimated in 1982 and 1989.[6]

Due to industry diversity, EPA developed three subcategories based on the similarity in the mills, raw materials used, products manufactured, production processes employed, mill size, age, and treatment costs. These subcategories are:

- Integrated Mills;
- Secondary Fibers Mills; and
- Nonintegrated Mills.

TABLE 2-5. QUANTITIES OF WASTES MANAGED IN SUBTITLE D SURFACE IMPOUNDMENTS

Industry	Quantity (Dry Metric Tons)
Electric power generation (SIC 4911)	28,497,800
Fertilizer and other agricultural chemicals (SIC 2873-2879)	8,640,800
Food and kindred chemicals (SIC 20)	NA
Industrial organic chemicals (SIC 2819)	38,058,700
Leather and leather products (SIC 31)	1,200
Machinery, except electrical (SIC 35)	NA
Pulp and paper (SIC 26)	579,700
Petroleum refining industry (SIC 29)	NA
Pharmaceutical preparations (SIC 2834)	NA
Plastics and resins manufacturing (SIC 2821)	30,513,700
Primary iron and steel manufacturing and ferrous foundries (SIC 3312-3321)	14,563,000
Primary non-ferrous metals manufacturing and Non-ferrous foundries (SIC 3330-3399)	147,300
Totals:	121,002,200

TABLE 2-6. PERCENT OF SUBTITLE D SURFACE IMPOUNDMENTS ACCEPTING INDUSTRIAL WASTE BY ACREAGE CATEGORY

Acreage Category (Acres)	Percent of Impoundments (%)
<0.1	10.8
0.1-0.4	24.8
0.5-0.9	33.6
1-5	17.0
6-10	7.0
11-100	5.8
>100	1.0

TABLE 2-7. NUMBER OF SMALL QUANTITY GENERATORS BY INDUSTRY GROUP AND QUANTITY OF WASTE GENERATED[13]

Industry	Number of Small Quantity Generators	Waste Quantity (MT/yr)[a]
Pesticide end users	1,623	1,122
Pesticide-application services	9,444	8,444
Chemical manufacturing	753	2,373
Wood preserving	193	715
Formulators	902	2,333
Laundries	15,646	13,418
Other services	16,322	10,706
Photography	9,355	18,052
Textile manufacturing	272	650
Vehicle maintenance	224,632	427,287
Equipment repair	1,795	943
Metal manufacturing	37,320	64,652
Construction	12,677	5,033
Motor freight terminals	148	161
Furniture/wood manufacture and refinishing	3,355	3,703
Printing/ceramics	24,640	18,307
Cleaning agents and cosmetic manufacturing	543	1,569
Other manufacturing	2,564	5,361

(CONTINUED)

TABLE 2-7. (CONTINUED)

Industry	Number of Small Quantity Generators	Waste Quantity (MT/yr)
Paper industry	181	544
Analytical and clinical laboratories	6,409	7,171
Educational and vocational establishments	1,179	N/R[a]
Wholesale and retail establishments	5,731	3,876
Total:	377,981	597,625

[a]Metric tons per year
[b]Not Reported

Integrated mills manufacture paper products or market pulp from wood which is prepared, pulped, and bleached on-site. Some pulp may be purchased for blending with pulp produced on-site to achieve the desired paper properties. Nonintegrated mills manufacture paper products by blending purchased pulps to achieve the desired paper properties. The Secondary Fibers mills get their major fiber source from purchased wastepaper. Wastepaper is mildly cooked, bleached (if necessary) and possibly blended with purchased pulp to achieve desired paper properties.

The majority of the organics are formed in the pulping and bleaching of virgin pulp. For this reason, the integrated pulp and paper mills are most likely to generate waste streams with high organic loadings. Secondary fibers mills and non-integrated mills do not generate wastewater with concentrations of organics as high as the streams generated in integrated mills.

Based on EPA data, approximately 49 percent of the pulp and paper mills are direct dischargers; 34 percent are indirect dischargers; two percent use a combination of both; and seven percent are zero dischargers. Eight percent of the pulp and paper mills did not report their method of discharge. The volume of wastewater discharged averages about 2.8 MGD per facility.

During the pulping process, the lignin present in the wood is broken down into simpler organic compounds such as methanol and acetone. These soluble organics are washed from the pulp and are concentrated in the spent pulping liquor. In the recovery process of this pulping liquor, the organics are evaporated and condensed. The resulting condensate streams are rich in organics and are sometimes sewered without treatment. Organics are also formed as additional lignin breaks down in the bleaching stages. In the presence of chlorine, chloroform and other chlorinated organics are formed and are washed from the pulp. These organics are readily volatilized from the bleach plant wash waters. Digester vent condensates, evaporator condensates, and bleach plant wash waters may contain high organic loadings. Facilities visited in 1989 by EPA are using air strippers and steam strippers to lower organics concentration and total reduced sulfur from their condensate streams, but some condensate streams are still sewered. Typically, wastewaters discharged from processes in the pulp and paper industry are discharged into grated trenches at elevated temperatures. Any controls in place that control organics have been put in place to control total reduced sulfur emissions and the resulting odor. No attempts to suppress emissions of volatile organics from the wastewater collection and treatment systems were noted.

TABLE 2-8. MILL POPULATION

Subcategory	Total Number (1982)	Estimate (1989)
Market kraft	13	14
Dissolving kraft	3	3
BCT kraft	9	8
Alkaline fine	20	24
Unbleached kraft & semi-chemical	11	8
Unbleached kraft - liner	17	21
Unbleached kraft - bag	12	5
Semi-chemical	20	16
Dissolving sulfite	6	6
Papergrade sulfite	16	11
Groundwood CMN	6	5
Groundwood fine	9	9
Groundwood TMP	4	7
Deink - fine	5	5
Deink - news	4	4
Deink - tissue	16	21
Tissue from wastepaper	23	19
Wastepaper-molded product	20	13
Paperboard from wastepaper	161	132
Builders' papers & roofing felts	66	21
NI - fine	32	35
NI - tissue	29	22
NI - lightweight	12	10
NI - electrical	6	4
NI - fine cotton	7	6
NI - filter nonwoven	14	13
NI - board	16	12
Misc. - integrated	88	91
Misc. - nonintegrated	34	38
Misc. - secondary fibers	16	20
Total:	695	603

The pulp and paper industry has not been included in this document because of the characteristics of the wastewater and the lack of data. However, in the future, this industry may be added to the list of industries required to control wastewaters containing organic compounds. The quantity of wastewater generated in a typical pulp and paper facility is 2.8 MGD as opposed to an average of 0.78 MGD in the OCPSF industry. Available data show lower concentrations of organics in pulp and paper wastewaters than are found in the four industries covered by this document. The wastewaters in the pulp and paper industry also typically have higher total suspended solids concentrations and pH values above 11 or below 3. These characteristics make the pulp and paper wastewaters less amenable to steam stripping with carbon steel equipment.

2.6 REFERENCES

1. Office of Water Regulations and Standards. U. S. Environmental Protection Agency. Washington, D. C. Development for Effluent Limitations Guidelines and Standards for the Organic Chemicals and Plastics and Synthetic Fibers Point Source Category. Volume I (BPT). EPA 440/1-83/009b (NTIS PB83-205633), February 1983. pp. 11-39, 94-96, 105-167, 195-198.

2. Office of Water Regulations and Standards. U. S. Environmental Protection Agency. Washington, D. C. Development Document for Effluent Limitations Guidelines and Standards for the Organic Chemicals and Plastics and Synthetic Fibers Point Source Category Volume II (BAT). EPA 440/1-83/009b (NTIS PB83-205641), February 1983. pp. II-7 - II-14, III-1 - III-9, IV-10 - IV-78, V-1 - V-15, VII-1 - VII-22.

3. Office of Water Regulations and Standards. U. S. Environmental Protection Agency. Washington, D. C. Development for Effluent Limitations Guidelines and Standards for the Pesticide Point Source Category. EPA 440/1-85/079 (NTIS PB86-150042/REB), October 1985. pp. III-9 - III-16, VI-4 - VI-47.

4. Office of Water. U. S. Environmental Protection Agency. Washington, D. C. Development Document for Final Effluent Limitations Guidelines, New Source Performance Standards and Pretreatment Standards for the Pharmaceutical Manufacturing Point Source Category. EPA 440/1-83/084 (NTIS PB84-180066/REB), September 1983. pp. 17-33, 49-75, 103-137.

5. Office of Water Regulations and Standards. U. S. Environmental Protection Agency. Washington, D. C. Development Document for Effluent Limitation Guidelines and Standards for the Pulp, Paper and Paperboard Builders Paper and Board Mills Point Source Category. EPA 440/1-82/025 (NTIS PB83-163949), October, 1982. pp. 77-87, 183-235, 363-385.

6. Meeting notes. Penny E. Lassiter, Environmental Protection Agency, Chemicals and Petroleum Branch, with Tom O'Farrell, Office of Water Regulations and Standards, Industrial Technology Division, March 3, 1989.

7. Office of Water Regulations and Standards. U. S. Environmental Protection Agency. Washington, D. C. Draft Decision Document for Hazardous Waste Treatment Industry Effluent Guidelines Development. EPA Contract No.: 68-01-6947, August 17, 1987.

8. Federal Register, October 4, 1985. Pesticide Effluent Limitations Guidelines, Notice of Availability. 50 FR 40674.

9. U. S. Environmental Protection Agency. Code of Federal Regulations. Organic Chemicals and Plastics and Synthetic Fibers; Point Source Category Effluent Limitations Guidelines Pretreatment Standards; and Standards of Performance for New Sources. Title 40, Chapter 1, Subchapter N, Parts 414 and 416. Washington, D. C. Office of the Federal Register. July 17, 1985. p. 29072.

10. Letter and attachments from Radian Corporation to Penny E. Lassiter, EPA/OAQPS, July 10, 1987, on the summary of Section 114 responses.

11. U. S. Environmental Protection Agency. Code of Federal Regulations. Pharmaceutical Manufacturing Point Source Category Effluent Limitations Guidelines, Pretreatment Standards, and New Source Performance Standards. Title 40, Chapter 1, Subchapter N, Part 439. Washington, D. C. Office of the Federal Register. October 27, 1983. p. 49809.

12 Office of Air Quality Planning and Standards. U. S. Environmental Protection Agency. Research Triangle Park, North Carolina. Control of Volatile Organic Emissions from Manufacture of Synthesized Pharmaceutical Products, OAQPS Guideline Series, EPA-450/2-78- 029 (NTIS PB290-580/0), December 1978. p 1-1 - 2-5.

13. Office of Solid Waste and Emergency Response. U. S. Environmental Protection Agency. Subtitle D Study Phase I Report. Washington, D. C. EPA 530-SW-86-054, October 1986.

14. U. S. Environmental Protection Agency. Code of Federal Regulations. Pulp, Paper, and Paperboard and the Builders' Paper and Board Mills Point Source Categories Effluent Limitations Guidelines, Pretreatment Standards, and New Source Performance Standards. Title 40, Chapter 1, Subchapter N, Parts 430 and 431. Washington, D. C. Office of the Federal Register. November 18, 1982. p. 52006.

3. VOC Emissions During Wastewater Collection and Treatment

In the manufacture of chemical products, wastewater streams are generated which contain organic compounds. These wastewaters are collected and treated in a variety of ways. Some of these collection and treatment steps result in the emission of VOC from the wastewater to the air. This chapter provides a discussion of the potential VOC emission sources and presents estimates of emissions for model systems. Wastewater sources are discussed in Section 3.1. Potential sources of VOC emissions during wastewater collection and treatment and factors affecting emissions from these sources are discussed in Section 3.2. Overall VOC emission estimates from three example waste stream collection and treatment systems are presented in Section 3.3. Development of emission factors are presented in Appendices A and B.

3.1 SOURCES OF ORGANIC-CONTAINING WASTEWATER

The industries discussed in Chapter 2 differ in structure and manufacture a wide variety of products. However, many of the chemical processes employed within these industries use similar organic compounds as raw materials, solvents, catalysts, and extractants. In addition, many of these processes also generate similar organic by-products during reaction steps. Consequently, many of the wastewater streams generated by the targeted industries are similar in organic content. These organic containing wastewater streams result from both the direct and indirect contact of water with organic compounds.

3.1.1 Direct Contact Wastewater

Water comes in direct contact with organic compounds due to many different chemical processing steps. As a result of this contact, wastewater streams are generated which must be discharged for treatment or disposal. A few sources of process wastewater are:

- Water used to wash impurities from organic products or reactants;
- Water used to cool or quench organic vapor streams;

- Condensed steam from jet eductor systems pulling vacuum on vessels containing organics;
- Water used as a carrier for catalysts and neutralizing agents (e.g., caustic solutions); and
- Water formed as a by-product during reaction steps.

Two additional types of direct contact wastewater are landfill leachate and water used in equipment washes and spill cleanups. These two types of wastewater are normally more variable in flow and concentration than the streams previously discussed. In addition, landfill leachate and spill cleanups may be collected for treatment differently than the wastewater streams discharged from process equipment such as scrubbers, decanters, evaporators, distillation columns, reactors, and mixing vessels.

3.1.2 Indirect Contact Wastewater

Wastewater streams which do not come in contact with organic compounds in the process equipment are defined as "indirect-contact" wastewater. However, a potential exists for organic contamination of these wastewater types. Non-contact wastewater may become contaminated as a result of leaks from heat exchangers, condensers, and pumps. These indirect contact wastewaters may or may not be collected and treated in the same manner as direct contact wastewaters. Pump seal water is normally collected in area drains which tie into the process wastewater collection system. This wastewater is then combined with direct contact wastewater and transported to the wastewater treatment plant. Wastewater contaminated from condenser and heat exchanger leaks are often collected in different systems and bypass some of the treatment steps used in the treatment plant. The organic content in these streams can be minimized by implementing an aggressive leak detection program.

3.2 SOURCES OF AIR EMISSIONS

Wastewater streams are collected and treated in a variety of ways. Generally, wastewater passes through a series of collection and treatment units before being discharged from a facility. Many of these collection and treatment system units are open to the atmosphere and allow organic-containing wastewaters to contact ambient air. Whenever this happens, there is a potential for VOC emissions. The organic pollutants volatilize in an attempt to exert their equilibrium partial pressure above the wastewater. In doing so, the organics are emitted to the ambient air surrounding the collection and treatment units. The magnitude of VOC emissions depends greatly on many factors such as the physical properties of the pollutants, the temperature of the wastewater, and the design of the individual collection and treatment units. All of these factors as well as the general scheme used to collect and treat facility wastewater have a major effect on VOC emissions.

Collection and treatment schemes are facility specific. The flow rate and organic composition of wastewater streams at a particular facility are functions of the processes used. The wastewater flow rate and composition, in turn, influence the sizes and types of collection and treatment units that must be employed at a given facility. Figure 3-1 illustrates a typical scheme for collecting and treating process wastewater generated at a facility and the opportunity for volatilization of organics. Figure 3-1 illustrates wastewater being discharged from a piece of process equipment into a drain. Drains are typically open to the atmosphere and provide an opportunity for volatilization of organics in the wastewater. The drain is normally connected to the process sewer line which carries the wastewater to the downstream collection and treatment units. Figure 3-1 illustrates the wastewater being carried past a manhole and on to a junction box where several process wastewater streams are joined. The manhole provides an escape route for organics volatilized in the sewer line. In addition, the junction box may also be open to the atmosphere, allowing organics to volatilize. Wastewater is discharged from the junction box to a lift station where it is pumped to the treatment system. The lift station is also likely to be open to the atmosphere, allowing volatilization of organics. The equalization basin, the first treatment unit shown in Figure 3-1, regulates the wastewater flow and pollutant compositions to the remaining treatment units. The equalization basin also typically provides a large area for wastewater contact with the ambient air. For this reason, emissions may be relatively high from this unit. Suspended solids are removed in the clarifier, and the wastewater then flows to the aeration basin where microorganisms act on the organic constituents. Both the clarifier and the aeration basin are typically open to the atmosphere. In addition, the aeration basin is normally aerated either mechanically or with diffused air.

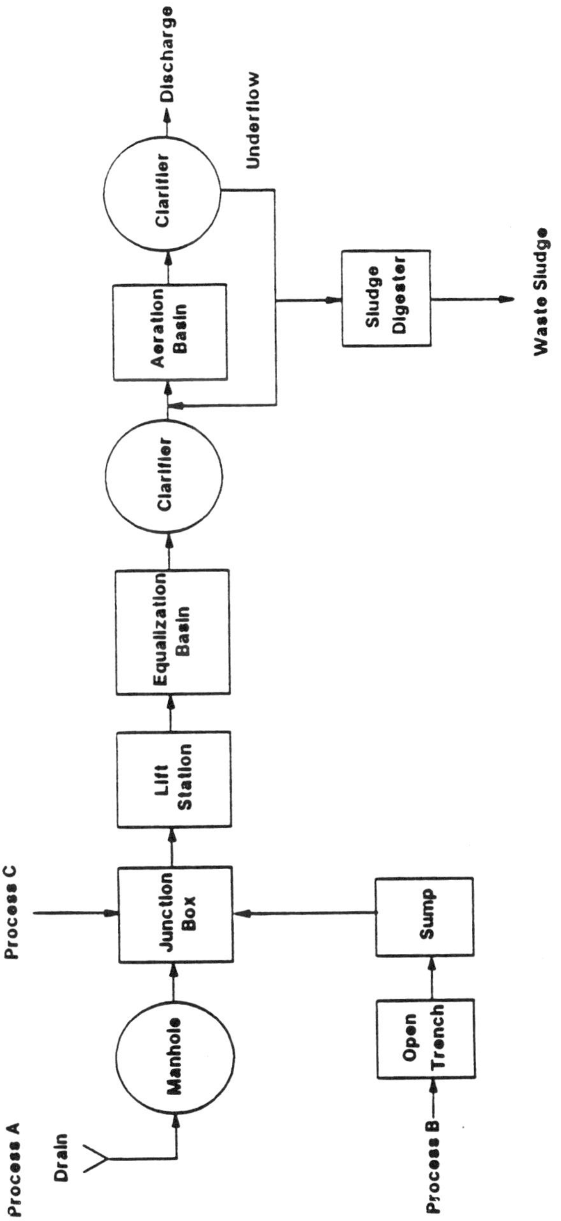

Figure 3-1. Typical wastewater collection and treatment scheme.

Wastewater leaving the aeration basin normally flows through a secondary clarifier for solids removal before it is discharged from the facility. The secondary clarifier is also likely to be open to the atmosphere. The solids which settle in the clarifier are discharged partly to a sludge digester and partly recycled to the aeration basin. Finally, waste sludge from the sludge digester is generally hauled off for landfilling or land treatment.

As mentioned previously, the types of components used to collect and treat wastewater are facility-specific. Figure 3-1 serves only as an example scheme for collecting and treating facility wastewater. Table 3-1 presents a more complete list of components that may be sources of emissions in facility collection and treatment systems. Although included in Figure 3-1, sludge digesters are not included in Table 3-1 since they are not considered to be major emission sources. The following sections will discuss each of the emission sources listed in Table 3-1. A diagram of each unit is provided including typical ranges of design parameters. The function of each unit, emission mechanisms, and factors affecting emissions are also discussed. Techniques for estimating emissions from each unit are presented in Appendix A or Appendix B.

3.2.1 Drains

Wastewater streams from various sources throughout a given process are normally introduced into the collection system through process drains. Individual drains are usually connected directly to the main process sewer line. However, they may also drain to trenches, sumps, or ditches. Some drains are dedicated to a single piece of equipment such as a scrubber, decanter, or stripper. Others serve several sources. These types of drains are located centrally between the pieces of equipment they serve and are referred to as area drains.

Many of the drains discussed above are open to the atmosphere. That is, they are not equipped with a water seal pot or p-trap to prevent emissions of organic vapors. A typical open drain configuration is shown in Figure 3-2, and the typical range of dimensions are listed. As shown in Figure 3-2, a straight section of pipe, referred to as the drain riser, extends vertically from the main process sewer line to just above ground level. A process drain line introduces wastewater to the mouth of the drain.

TABLE 3-1. EMISSION SOURCES IN WASTEWATER COLLECTION AND TREATMENT SYSTEMS

<u>Wastewater Collection System:</u>
Drains
Junction Boxes
Lift Stations
Manholes
Trenches
Sumps
Surface Impoundments
<u>Wastewater Treatment Units:</u>
Weirs
Oil/Water Separators
Equalization Basins or Neutralization Basins
Clarifiers
Aeration Basins
pH Adjustment Tanks
Flocculation Tanks
Surface Impoundments

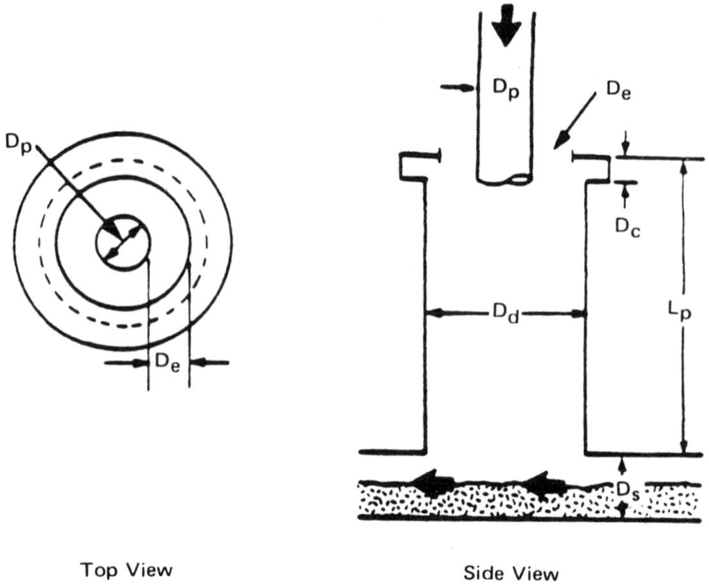

Design Parameter	Range	Typical Design
Drain riser height, L_p (m)	0.3 - 1.2	0.6
Drain riser diameter, D_d (m)	0.1 - 0.3	0.2
Process drain pipe diameter, D_p (m)	0.005 - 0.15	0.1
Effective diameter of drain riser, D_e (m)	0.005 - 0.15	0.1
Drain riser cap thickness, D_c (cm)	0.5 - 0.7	0.6
Sewer Diameter, D_s (m)	0.6 - 1.2	0.9

Figure A-1. Typical drain configuration.

3.2.1.1 VOC Emission Mechanisms

Emissions occur from drains by diffusive and convective mechanisms.[1] As wastewater flows through the drain, organics volatilize in an attempt to reach equilibrium between the aqueous and vapor phases. The organic vapor concentration in the headspace at the bottom of the drain riser is much higher than ambient concentrations. Due to this concentration gradient, organics diffuse from the drain into ambient air through the opening at the top of the drain riser. In addition, if the temperature of the wastewater flowing through the sewer is greater than the ambient air temperature, this temperature gradient will induce air flow from the vapor headspace in the sewer line. This air flow passes through the drain riser and into the ambient air. The convective forces created by this air flow establishes convective mass transfer of the organics. Air flows resulting from wind blowing over or into the drain, or from wind currents entering another sewer opening and flowing through the sewer, also aid the mass transfer.

3.2.1.2. Factors Affecting Emissions from Drains. Drain emission rates are affected by a number of factors. These factors include the composition and physical properties of the organics in wastewater entering the drain and flowing through the sewer line below the drain, the temperature of the wastewater, drain design characteristics, and climatic factors.[2] The volatility of the organics in water is the most significant physical property affecting the rate of emissions from drains. The Henry's Law constant (H) for an organic compound provides an indication of this physical property. Values for H are determined by measuring the equilibrium concentrations of an organic compound in the vapor and aqueous phases. However, the organic compound's vapor pressure and water solubility are sometimes used, when laboratory data are not available, to estimate values for H. Using these data, the value for H is estimated by computing the ratio of the organic compound's vapor pressure to its water solubility at the same temperature. Organic compounds with low water solubilities and high vapor pressures exhibit the highest values for H and, therefore, these compounds tend to volatilize into the vapor phase more readily. Because the temperature of the wastewater affects the Henry's Law constant, this parameter will affect emissions.

Drain design characteristics also affect emissions. Drain design is dependent on the flow rate of the wastewater stream. The diameter of the drain riser must be large enough to prevent the wastewater from overflowing on to the ground. As the diameter increases, so does the surface area exposed to ambient air. This increase in surface area increases the rate of air emissions. The length of the drain riser from the mouth of the drain to the process sewer is another design parameter which affects emissions. Pollutants are more readily emitted to the atmosphere from a short drain riser due to the smaller resistance to diffusional and convective mass transfer.

Emission rates from a drain are also affected by climatic factors. These include ambient air temperature and wind speed and direction. Differences in temperature between the ambient air and the vapors at the bottom of the drain establish pressure and density gradients across the drain riser. These gradients generate bulk vapor flow from the sewer headspace to the top of the drain riser. This bulk flow increases convective mass transfer of organic compounds to the air surrounding the drain. Wind speed has a similar effect. As wind moves across the top of the drain riser, it creates an aspirator-like effect. The lower pressure at the mouth of the drain "pulls" vapors from the sewer line headspace at the bottom of the drain riser. This pressure gradient, therefore, increases the convective mass transfer of organic compounds to air surrounding the drain. Wind blowing into any upstream opening will also increase the volatilization rate of the organics. Development of the drain emission factor is presented in Appendix B.

3.2.2 Manholes

Manholes are service entrances into process sewer lines which permit inspection and cleaning of the sewer line. They are normally placed at periodic lengths along the sewer line. They may also be located where sewers intersect or where there is a significant change in direction, grade, or sewer line diameter. Figure 3-3 illustrates a typical manhole configuration, and presents typical ranges of manhole dimensions. The lower portion of the manhole is usually cylindrical, with a typical inside diameter of four feet to allow adequate space for workmen. The upper portion tapers to the diameter of the opening at ground level. The opening is normally about two feet in diameter and covered with a heavy cast-iron plate. The cover usually contains two to four holes for ventilation so that the manhole cover can be grasped for removal.

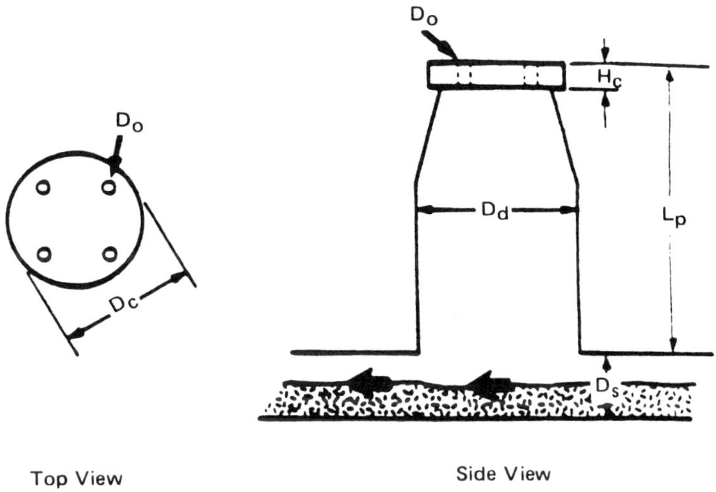

Design Parameter	Range	Typical Design
Manhole diameter, D_d (m)	0.6 - 1.8	1.2
Manhole height, L_p (m)	0.3 - 1.8	1.2
Manhole cover diameter, D_c (m)	0.4 - 0.7	0.6
Diameter of holes in cover, D_o (cm)	1.2 - 3.8	2.5
Manhole cover thickness, H_c (cm)	0.5 - 0.7	0.6
Sewer Diameter, D_s (m)	0.6 - 1.2	0.9

Figure A-2. Typical Manhole Configuration.

3.2.2.1 VOC Emissions Mechanisms. Emissions occur from manholes by diffusive and convective mechanisms.[3] As wastewater moves through the sewer lines, organics volatilize in an attempt to reach equilibrium between the aqueous phase and the vapor headspace in the sewer line. The organic vapor concentration in the headspace above the wastewater is much higher than the concentration of organics in the ambient air above the manhole. Due to this concentration gradient, organics will diffuse from the sewer line into the ambient air through the manhole openings. In addition, if the temperature of the wastewater flowing through the sewer is greater than the ambient air temperature, this temperature gradient will induce air flow from the vapor headspace in the sewer line. Wind entering through any opening in the sewer system may also create air flows in the sewer line. This air flow passes through the manhole openings and into the ambient air. The turbulence created by this air flow establishes convective mass transfer of the organics and increases the emission rate from the manhole.

3.2.2.2 Factors Affecting Emissions from Manholes. Emission rates from manholes are affected by the following types of factors: characteristics of the wastewater passing through the sewer line below the manhole, manhole design characteristics, and climatic factors.[4] Wastewater characteristics affecting emission rates include wastewater composition and temperature. Both the concentration and physical properties of the specific organic compounds present in the wastewater affect the emissions. As previously discussed for drains, air emissions are higher for organics with greater volatility in water and higher diffusivity in air. Wastewater temperature affects the volatility of the compound in water and, therefore, also affects emissions. This effect can be evaluated by measuring the change in the organic compound's Henry's Law constant with temperature.

Manhole design characteristics that affect emission rates are: the manhole diameter, length from the manhole cover down to the sewer line, the thickness of the manhole cover, and the number and diameter of the vent holes in the manhole cover. The length from the manhole cover to the sewer line is the distance organics must diffuse from the wastewater before being emitted to the ambient air. Therefore, an increase in this length will decrease the emission rate. The thickness of the cover adds to this diffusional length. The diameter of holes in the cover along with the number of holes determine the ultimate surface area available for diffusion and convection of organics into the ambient air.

Also similar to drains, climatic factors affecting emission rates are wind speed and direction and ambient air temperature. As previously discussed, wind speed will establish an aspirator-type effect at the cover of the manhole and increase convective mass transfer of the organics into the air. In addition, differences in temperature between the ambient air and the headspace air in the sewer line will establish density and thus pressure gradients between these two locations. These gradients establish flow patterns as previously discussed. This bulk air flow transports organic compounds from the wastewater surface to the ambient air, thereby increasing convective mass transfer. Wind blowing into the sewer system will also aid in the volatilization of the organics and will increase the emission rate from the sewer system components. Development of manhole emission factors are presented in Appendix B.

3.2.3 Junction Boxes

A junction box normally serves several process sewer lines. Process lines meet at the junction box to combine the multiple wastewater streams into one stream which flows downstream from the junction box. Generally, the flow rate is controlled by the liquid level in the junction box. Junction boxes are normally either square or rectangular and are sized based on the flow rate of the entering streams. As shown in Figure 3-4, uncontrolled junction boxes are open to the atmosphere. The range of typical junction box dimensions are provided at the bottom of Figure 3-4.

3.2.3.1 VOC Emission Mechanisms. Emissions occur from junction boxes predominantly by convective mass transfer. Organics in the wastewater volatilize into the ambient air just above the liquid surface in an attempt to reach equilibrium between the liquid and vapor phases. Since the organic vapors above the liquid are in contact with the ambient air, these organic vapors can be swept into the atmosphere by wind blowing across the top of the junction box.

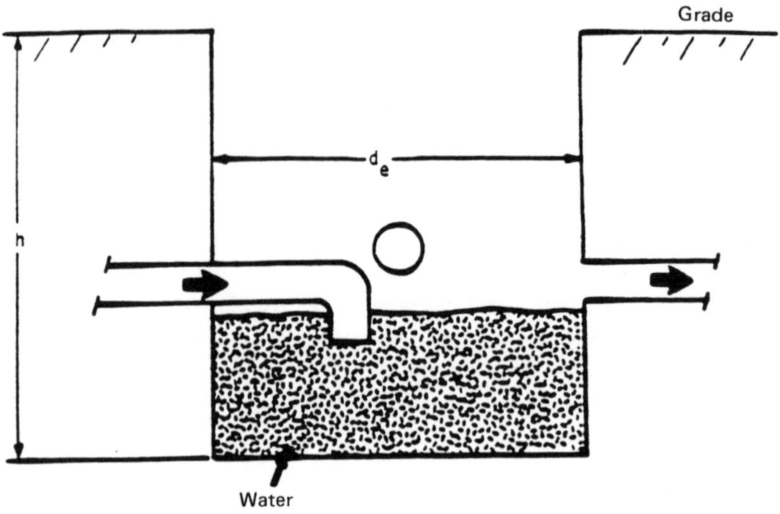

Design Parameter	Range	Typical Design
Effective diameter, d_e (m)	0.3 - 1.8	0.9
Grade height, h (m)	1.2 - 1.8	1.5
Water Depth (m)	0.6 - 1.2	0.9
Surface area (m^2)	0.007 - 2.5	0.7

Figure 3-4. Typical Junction Box Configuration.

3.2.3.2. Factors Affecting Emissions from Junction Boxes. Emission rates from junction boxes are affected by the following types of factors: characteristics of the wastewater flowing through the junction box, design of the junction box, and climatic factors.[5] Wastewater characteristics such as the concentration and physical properties of the specific organic compounds present in the wastewater have a significant effect on air emissions. Increases in organic compound concentration and physical properties such as the compound's volatility in water increase air emission rates. Higher wastewater temperatures also increase the organic compound's volatility in water as previously discussed for drains. Therefore, an increase in temperature will increase the emission rate.

Junction box design characteristics that affect emissions are: the the fetch to depth ratio, the water turbulence in the junction box, and the liquid surface area. Fetch is defined as the linear distance across the junction box in the direction of the wind flow. Depth is represented by the average liquid level in the junction box. As the liquid depth in the junction box increases, so does the resistance to liquid phase mass transfer. That is, organic compounds must overcome more resistance before they reach the water surface. Once these organics reach the surface, the fetch length provides the route for volatilization into the ambient air. Therefore, increases in the fetch to depth ratio for the junction box increase air emissions.

Water turbulence enhances liquid phase mass transfer.[6] In completely smooth flow through the junction boxes, pollutants slowly diffuse to the water surface to replace the volatilizing pollutants. In turbulent flow through the junction box, the organic compounds are carried much more rapidly to the surface by the turbulent water. Therefore, more organic compounds are exposed to the surface air, and the emission rate is increased. If the sewer lines feed water to the junction box above the liquid surface, the exposure of organic compounds to the surface air is also increased. The water spills into the junction box causing splashing and additional turbulence at the liquid surface which increases emissions. In addition, wind entering the sewer system through an upstream component may exit the junction box saturated with organics. These effects can be minimized by introducing water to the junction box below the liquid surface. The final design characteristic affecting emissions is surface area. An increase in surface area at constant depth increases the hydraulic (water) retention time in the junction box. Therefore, not only is the area for volatilization increased but so is the time available for volatilization to occur.

Ambient wind speed is a climatic factor affecting air emissions. As the wind speed increases, so does convective mass transfer due to the additional air turbulence above the wastewater surface. This wind speed effect is more prominent when the liquid level is closer to the top of the junction box. If the sewer lines feed the junction box above the liquid surface, wind blowing into the sewer system through an upstream component may also have an effect on emissions from the junction box. Development of junction box emission factors are presented in Appendix A. Appendix B also presents the development of emission factors for junction boxes. However, the emission estimates presented in Appendix B are based on the assumption that the major emissions from a junction box are a result of air leaving the sewer line through the junction box, in equilibrium with the water in the sewer. Appendix A presents an emission estimate based on the assumption that evaporation from the junction box, assisted by turbulence caused by influent streams, is the primary cause of emissions.

3.2.4 Lift Station

Lift stations are usually the last collection unit prior to the treatment system, accepting wastewater from one or several sewer lines. The main function of the lift station is to provide sufficient head pressure to transport the collected wastewater to the treatment system. A pump is used to provide the head pressure and is generally designed to operate on and off based on preset high and low liquid levels. An open top lift station is illustrated in Figure 3-5. As shown in Figure 3-5, lift stations are usually rectangular in shape and greater in depth than length or width.

3.2.4.1 VOC Emission Mechanisms. Emissions occur from lift stations predominantly by convective mass transfer. Organics in the wastewater volatilize into the ambient air just above the liquid surface in an attempt to reach equilibrium between the liquid and vapor phases. Since the organic vapors above the liquid are in contact with the ambient air, these organic vapors can be swept into the atmosphere by wind blowing across the top of the lift station.

3.2.4.2 Factors Affecting Emissions from Lift Stations. The factors affecting emissions from lift stations are similar to the factors affecting

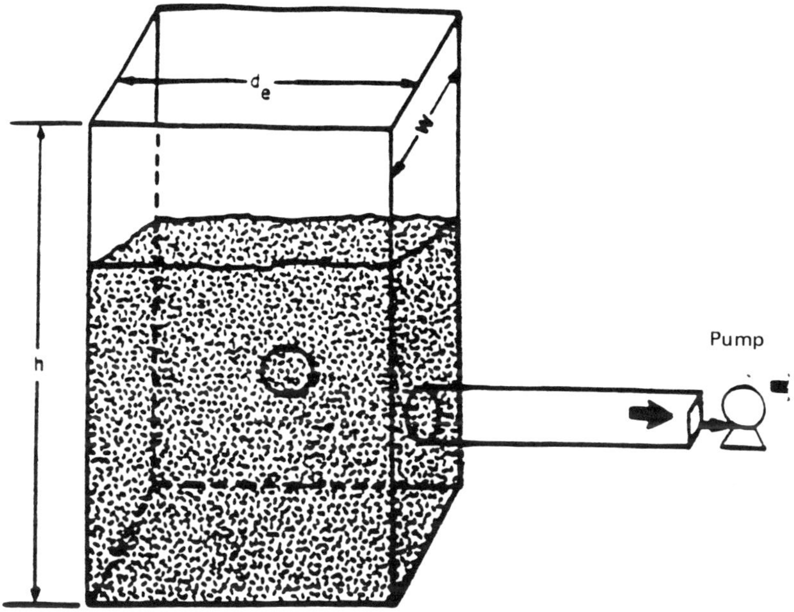

Design Parameter	Range	Typical Design
Effective diameter, d_e (m)	1.2 - 3.0	1.5
Width, W (m)	1.4 - 3.6	1.8
Grade height, h (m)	1.8 - 2.4	2.1
Water Depth (m)	1.2 - 1.8	1.5
Surface area (m^2)	1.1 - 7.1	1.8

Figure 3-5. Typical Lift Station Configuration.

emissions from junction boxes discussed in Section 3.2.3.2. These factors are: the concentration and physical properties of the organics present in the wastewater, lift station design characteristics, and climatic factors.[7] Increases in organic compound concentration and volatility in water increase the rate of emissions from lift stations. Because increases in temperature increase the volatility of many compounds, warmer wastewater temperature will increase the emissions rate of most organic compounds.

The design characteristics which affect air emission rates from lift stations are: the liquid surface area, the water turbulence in the lift station, and the fetch to depth ratio. Increases in these design parameters will increase air emission rates. The hydraulic retention time which is a function of wastewater flow rate and the liquid volume in the lift station also has an effect on emissions. An increase in retention time will result in an increase in air emissions due to the additional time for volatilization.

In addition to these design parameters, operation of the lift station affects air emissions. The liquid level in a lift station normally rises and falls based on the wastewater flow to the unit. As the level rises, the wastewater acts as a piston displacing organic vapors above the liquid surface into the ambient air. The linear distance between the low and high level limits in the lift station determine the amount of displacement during normal operation. As this distance increases, displacement increases and so does the emission rate. Also, at lower liquid levels, wastewater is normally spilling into the lift station above the liquid surface. This causes an increase in turbulence which increases liquid phase mass transfer. Therefore, volatilization occurs more rapidly above the surface of the rising liquid. At the higher liquid levels, the sewer lines feeding the lift station are often submerged which reduces splashing above the liquid surface. Development of lift station emission factors is presented in Appendix A.

3.2.5 Trench

Trenches are normally used to transport wastewater from the point of process equipment discharge to subsequent wastewater collection units such as junction boxes and lift stations. This mode of transport replaces the drain scenario as a method for introducing process wastewater into the downstream collection system. In older plants, trenches are often the primary mode of wastewater transportation in the collection system. Trenches are often interconnected throughout the process area and handle pad water runoff, water from equipment washes and spill cleanups, as well as process wastewater discharges. Normally, the length of the trench is determined by the general locations of the process equipment and the downstream collection system units. This length typically ranges from 50 to 500 feet. Trench depth and width are dictated by the wastewater flow rate discharged from process equipment. The depth and width of the trench must be sufficient to accommodate expected as well as emergency wastewater flows from the process equipment. Figure 3-6 illustrates a typical trench configuration.

3.2.5.1 VOC Emission Mechanisms. Emissions from trenches, like junction boxes and lift stations, occur predominantly by convective mass transfer. As wastewater flows through the trench, organic compounds volatilize into the ambient air above the liquid surface in an attempt to reach equilibrium between the liquid and vapor phases. Since the organic vapors above the liquid are in contact with the ambient air, the organic vapors above the liquid can be swept into the atmosphere by wind blowing across the surface of the trench. Due to this volatilization, the organic compound concentration decreases as the wastewater flows through the trench. Therefore, the volatilization rate decreases somewhat as the wastewater moves downstream from the point of process equipment discharge.

3.2.5.2 Factors Affecting Emissions from a Trench. The factors which affect emissions from trenches are: the concentration and physical properties of the compounds in the wastewater, trench design characteristics, and climatic factors.[8] The effect of organic compound concentration and physical properties on air emissions from trenches are similar to the effect of these factors on emissions from well mixed, flow through impoundments (e.g., junction boxes, lift stations). Increases in organic compound concentration, volatility in water, and water temperature will increase air emissions. Wind speed also similarly affects air emissions from trenches. Increases in wind speed accelerate air emissions by increasing convective mass transfer.

The trench design characteristics which affect emission rate include the depth and width of the trench and the hydraulic retention time. Mass transfer rates increase as the depth of the trench becomes more shallow and the width of the trench becomes wider. The hydraulic retention time in the trench is a function of the wastewater flow rate and the volume of the trench. Longer

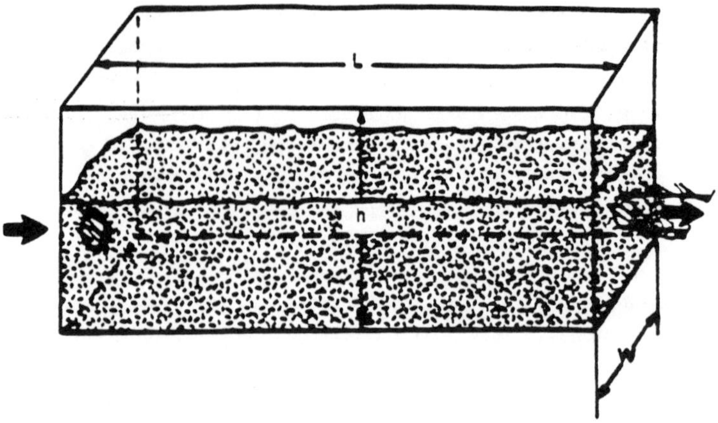

Design Parameter	Range	Typical Design
Trench length, L (m)	15 -150	15.2
Water depth (m)	0.3 - 0.9	0.6
Trench depth, h (m)	0.4 - 1.2	0.8
Trench width, W (m)	0.3 - 0.9	0.6

Figure 3-6. Typical trench configuration.

trenches increase the hydraulic retention for mass transfer to take place and, therefore, will increase air emissions. The grade (slope) of the trench is also important. Grade will have an effect on the turbulence of the wastewater flowing through the trench. An increase in turbulence will reduce the liquid phase resistance to mass transfer and thus increase air emissions. Development of trench emission factors is presented in Appendix B.

3.2.6 Sumps

Sumps are typically used for collection and equalization of wastewater flow from trenches prior to treatment. They are usually quiescent and open to the atmosphere. Typical diameters and depths are approximately 1.5 meters.

3.2.6.1 VOC Emission Mechanism. Emissions occur from sumps by both diffusive and convective mechanisms. As wastewater flows slowly through the sump, organics diffuse through the water to the liquid surface. These organics volatilize into the ambient air above the liquid, and can be swept into the air by wind blowing across the surface of the sump.

3.2.6.2 Factors Affecting Emissions from a Sump. The factors affecting emissions from a sump are similar to the factors affecting emissions from an equalization basin. These factors are: wastewater characteristics, wind speed, and sump design characteristics. The effects of wastewater characteristics and wind speed were previously discussed in the sections concerning junction boxes and lift stations. These two factors will have similar effects on the rate of air emissions from sumps. The design characteristics which affect air emission rates from sumps are: the fetch to depth ratio, the liquid surface area, and the hydraulic retention time. Fetch to depth ratios vary widely for different sumps. As the fetch to depth ratio increases, so does the mass transfer rate of organics into the ambient air. The hydraulic retention time, which is a function of the wastewater flow rate and volume of the sump, also has an effect on emissions. An increase in retention time provides additional time for organic compound volatilization to occur and, therefore, emissions increase. Likewise, an increase in the surface area of the sump increases the emissions rate. Appendix B also presents the development of emission factors for sumps. However, the emission estimates presented in Appendix B are based on the assumption that the major emissions from a sump are a result of density differences between the air leaving the sewer line through the sump and the surrounding air. Appendix A presents an emission estimate based on the assumption that diffusion from the water in the sump to the air is the primary cause of emissions.

3.2.7 Weir

Weirs act as dams in open channels. The weir face is normally aligned perpendicular to the bed and walls of the channel. Water from the channel normally overflows the weir but may pass through a notch, or opening, in the weir face. Because of this configuration, weirs provide some control of the level and flow rate through the channel. This control, however, may be insignificant compared to upstream factors which influence the supply of water to the channel. A typical weir configuration is illustrated in Figure 3-7.

3.2.7.1 VOC Emission Mechanism. As shown in Figure 3-7, often the water overflowing the weir proceeds down stair steps as shown in the figure. These stair steps serve to aerate the wastewater. The wastewater splashes off each step increasing the surface area of the water in contact with ambient air. This action increases diffusion of oxygen into the water which may be beneficial to the biodegradation process (often the next treatment step). However, this increased contact with air also accelerates emissions of volatile organics contained in the wastewater.[9,10] The organics volatilize from the surface of the falling water in an attempt to reach equilibrium between the liquid and vapor phases. The volatilizing organic compounds are swept into the ambient air surrounding the weir.

3.2.7.2 Factors Affecting Emissions from a Weir. The major factors affecting emissions from weirs are: wastewater characteristics, ambient wind speed, and weir design characteristics. The concentration and physical properties of the organic compounds in the wastewater have a significant effect on VOC emissions. The diffusivity in water of the specific organic compounds present in the wastewater may be the most significant physical property. Organics must first diffuse through the liquid phase before volatilizing from the surface of the falling wastewater. Therefore, an increase in organic compound diffusivity in water tends to increase the air emissions rate. Wastewater temperature affects diffusivity as well as other

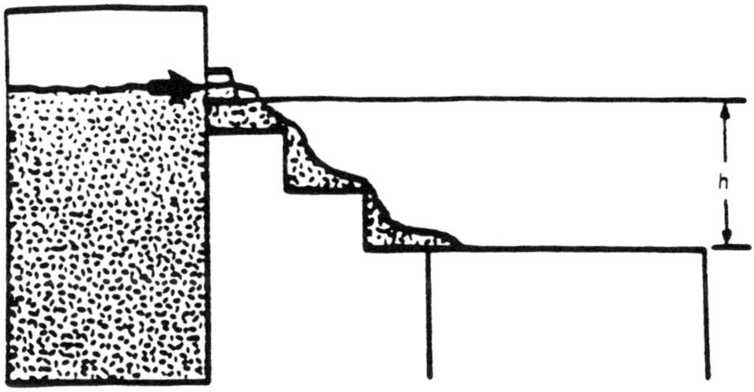

Design Parameter	Range	Typical Design
Weir height, h (m)	0.9 - 2.7	1.8

Figure 3-7. Typical weir configuration.

compound physical properties and, therefore, has an effect on air emissions.

Ambient wind speed has a significant effect on convective mass transfer. As the wastewater spills over the weir and splashes down the stair steps, increased liquid surface area is exposed. Wind sweeps the volatilized organics away from the liquid surface and carries them into the ambient air. As the wind speed increases, so does convective transfer of organics by this mechanism.

The height of the weir is the most significant design characteristic affecting emissions.[11] The height of the weir determines the length of time that the wastewater stream is falling through the air. Because this is the time period when the organics are being emitted to the air, an increase in weir height will increase the magnitude of air emissions. Development of weir emission factors is presented in Appendix A.

3.2.8 Oil/Water Separator

Oil/water separators are often the first step in the wastewater treatment plant but may also be found in the process area. The purpose of these units is to gravity separate and remove oils, scum, and solids contained in the wastewater. Most of the separation occurs as the wastewater stream passes through a quiescent zone in the unit. Oils and scum with specific gravities less than water float to the top of the aqueous phase. Heavier solids sink to the bottom. Most of the organics contained in the wastewater tend to partition to the oil phase. For this reason, most of these organic compounds are removed with the skimmed oil leaving the separator. The wastewater stream leaving the separator, therefore, is reduced in organic loading. Figure 3-8 illustrates a typical oil/water separator. The separator shown in the figure is open to the atmosphere.

3.2.8.1 VOC Emission Mechanism. Volatilization of organic compounds from the surface of an oil/water separator is a complex mass transfer phenomenon. Most organic compounds tend to partition to the oil phase which floats on the surface of the separator. The force behind volatilization is the drive to reach equilibrium between the concentration of organics in the oil layer and the vapor phase just above this layer. Organic compounds volatilizing into the vapor phase either diffuse or are swept by wind into the ambient air surrounding the oil/water separator.

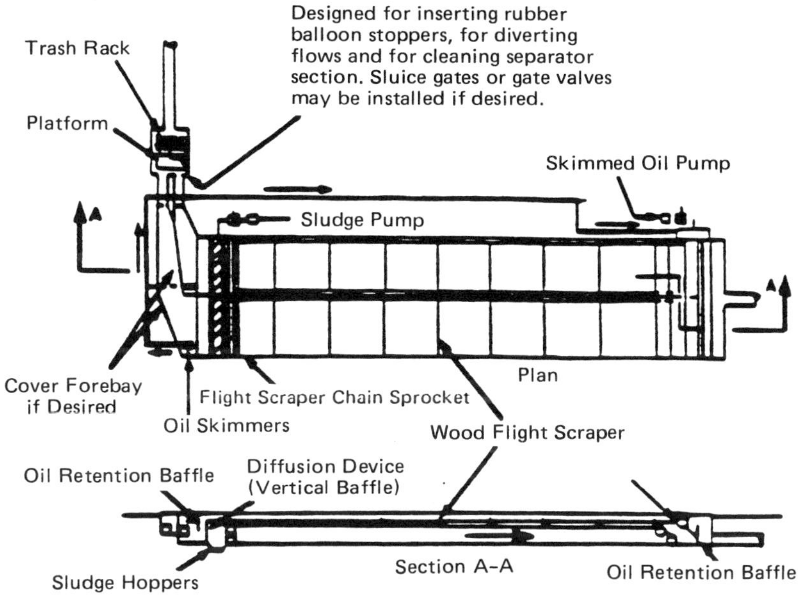

Design Parameter	Range	Typical Design
Separator length (m)	6.1 - 18.0	13.7
Separator width (m)	4.6 - 10.7	7.6
Retention time (hr)	0.6 - 1.0	0.8

Figure 3-8. Typical oil/water separator configuration.

3.2.8.2 Factors Affecting Emissions from an Oil/Water Separator. The factors affecting emissions from oil/water separators are: characteristics of the wastewater and oil layers, ambient wind speed, and design characteristics of the separator.[12] The concentration and physical properties of the organic compounds contained in the wastewater significantly affect emissions. The diffusivity of the organics in water and in the oil layer affect the mass transfer rate to the surface of the separator. Diffusivity of the organic compounds in air affect the rate of mass transfer into the ambient air.

The thickness of the oil layer also affects emissions. Organics that partition from the wastewater into the oil phase must diffuse through the oil layer to volatilize. If the separator is operated with a relatively thick nonvolatile oil layer, this layer may tend to act as a blanket suppressing emissions from the unit.

Ambient air speed above the oil surface increases convective mass transfer into the ambient air. Turbulence created by wind moving across the oil layer helps sweep organics into the ambient air. This effect is similar to the increase in convective mass transfer due to wind moving across well mixed, flow through units such as junction boxes and lift stations. The main difference is the existence of the oil layer which affects volatilization.

Design characteristics affecting emissions include the length and width of the oil/water separator. The length of the separator in the direction of the wind has a significant effect on the amount of convective mass transfer. This effect is similar to the fetch to depth ratio effect discussed for well mixed, flow through impoundments. An increase in the length of the separator increases area available for volatilization and, therefore, increases the emission rate. Development of oil/water separator emission factors is presented in Appendix A.

3.2.9 Equalization Basins

Equalization basins are used to reduce fluctuations in the wastewater flow rate and organic content to the downstream treatment processes. Equalization of wastewater flow rate results in more uniform effluent quality from downstream settling units such as clarifiers. Biological treatment performance can also benefit significantly from the damping of concentration and flow fluctuations. This damping protects biological processes from upset or failure due to shock loadings of toxic or treatment-inhibiting compounds.

Figure 3-9 illustrates a typical equalization basin. Normally, these basins use hydraulic retention time to ensure equalization of the wastewater effluent leaving the basin. However, some basins are equipped with mixers to enhance the equalization of organic compounds. Aerators may also be installed in some equalization basins to accelerate wastewater cooling or to saturate the wastewater with oxygen prior to secondary treatment.

3.2.9.1 VOC Emission Mechanisms. Emissions occur from equalization basins by both diffusive and convective mechanisms.[13] As wastewater flows slowly through the basin, organic compounds diffuse through the water to the liquid surface. These compounds volatilize into the ambient air above the liquid surface in an attempt to reach equilibrium between the liquid and vapor phases. Since the organic vapors above the liquid are in contact with the ambient air, these organic vapors can be swept into the air by wind blowing across the surface of the basin. If aerators are used in the basin, organic compounds are convectively transferred to the liquid surface. In addition, greater wastewater surface area is exposed to the wind and ambient area above the basin.

3.2.9.2 Factors Affecting Emissions from Equalization Basins. The factors affecting emissions from equalization basins are similar to the factors affecting emissions from other well mixed, flow through impoundments. These factors are: wastewater characteristics, wind speed, and equalization basin design characteristics.[14] The effect of wastewater characteristics and wind speed were previously discussed in the sections concerning junction boxes and lift stations. These two factors will have similar effects on the rate of air emissions from equalization basins. The design characteristics that affect air emission rates from equalization basins are: the fetch to depth ratio, the liquid surface area, the hydraulic retention time, and the degree of aeration. Fetch to depth ratios vary widely for different equalization basins. As the fetch to depth ratio increases, so does the mass transfer rate of organics into the ambient air. The hydraulic retention time, which is a function of the wastewater flow rate and volume of the basin, also has an effect on emissions. An increase in retention time provides additional time for organic compound volatilization to occur and, therefore, emissions increase. Likewise, an increase in the surface area of the basin increases the emissions rate. Also, if the basin is aerated, emissions will occur more

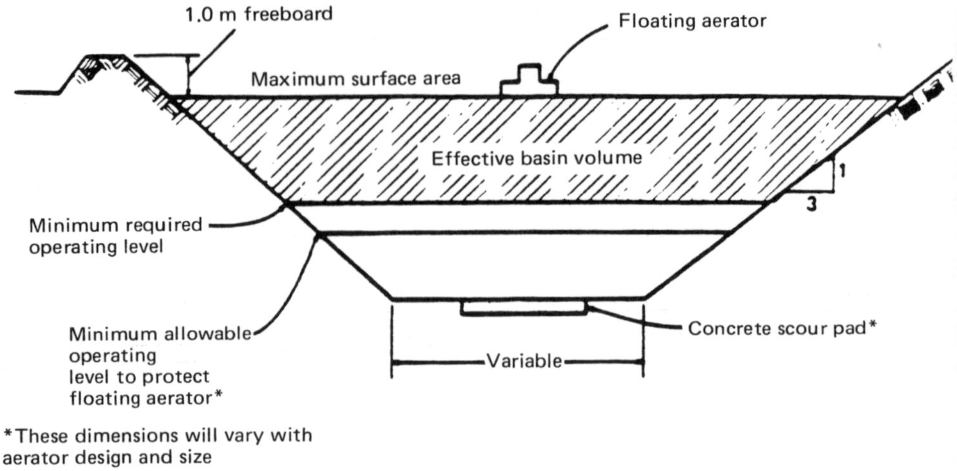

Design Parameter	Range	Typical Design
Effective Diameter (m)	20 -270	109
Surface Area (m^2)	300 - 57,000	9,290
Water depth (m)	1 - 8	2.9
Retention time (days)	0.2 - 20	5

Figure 3-9. Typical equalization basin.

rapidly from the aerated surface area. Therefore, as the degree of aeration (and aerated surface area) increases, so do air emissions. Development of equalization basin emission factors (aerated and nonaerated) is presented in Appendix A.

3.2.10 Clarifiers

The primary purpose of a clarifier is to separate any oils, grease, scum, and solids contained in the wastewater. Most clarifiers are equipped with surface skimmers to clear the water of floating oil deposits and scum. Clarifiers also have sludge raking arms which prevent accumulation of organic solids collected at the bottom of the tank.[15]

Figure 3-10 illustrates a typical clarifier. Clarifiers are generally cylindrical in shape. The depth and cross-sectional area of a clarifier are functions of the settling rate of the suspended solids and the thickening characteristics of the sludge. Clarifiers are designed to provide sufficient retention time for the settling and thickening of these solids.

3.2.10.1 VOC Emission Mechanism. Emissions occur from clarifiers by both diffusive and convective mechanisms.[16] As wastewater flows slowly through the clarifier, organic compounds diffuse through the water to the liquid surface. These compounds volatilize into the ambient air above the liquid surface in an attempt to reach equilibrium between liquid and vapor phases. Since the organic vapors above the liquid are in contact with the ambient air, these organic vapors can be swept into the air by wind blowing across the surface of the clarifier. In addition, clarifiers are often equipped with overflow weirs. These weirs provide additional contact with ambient air as the wastewater flows over the weir and spills into the effluent collection area.

3.2.10.2 Factors Affecting Emissions from a Clarifier. The factors affecting emissions from a clarifier are similar to the factors affecting emissions from other well mixed, flow through impoundments.[17] These factors are: the wastewater characteristics, wind speed, and clarifier design characteristics. Increases in wastewater temperature, organic concentration and organic compound physical properties such as volatility and diffusivity in water increase air emission rates. Likewise, increases in wind speed improve

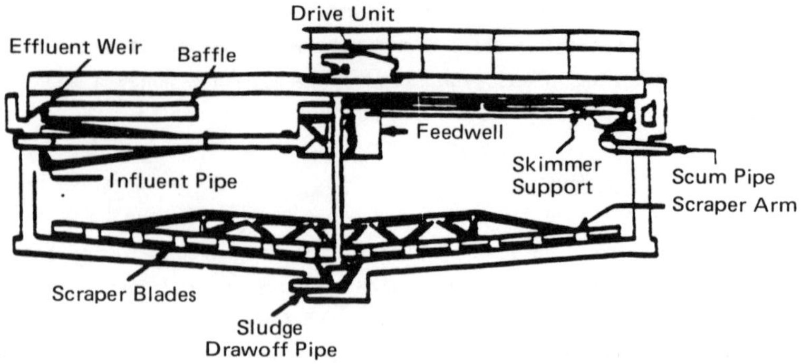

Design Parameter	Range	Typical Design
Diameter (m)	6.1 - 30.5	18.3
Depth (m)	2.4 - 4.5	3.5
Retention time (hr)	1.5 - 7	4.0

Figure 3-10. Typical clarifier configuration.

convective mass transfer and, therefore, also increase air emissions. The design characteristics which affect emission rates from clarifiers are the liquid surface area, the fetch to depth ratio, and the hydraulic retention time. Increases in the magnitude of these design parameters increase air emission rates. Increases in the hydraulic retention time in the clarifier also increase emissions. Development of clarifier emission factors is presented in Appendix A.

3.2.11 Biological Treatment Basins

Biological waste treatment is normally accomplished through the use of aeration basins. Microorganisms require oxygen to carry out the biodegradation of organic compounds which results in energy and biomass production. The aerobic environment in the basin is normally achieved by the use of diffused or mechanical aeration. This aeration also serves to maintain the biomass in a well mixed regime. The goal is to maintain the biomass concentration at a level where the treatment is efficiently optimized and proper growth kinetics are induced.

The performance of aeration basins is particularly affected by (1) mass of organic per unit area, (2) temperature and wind patterns, (3) hydraulic retention time, (4) dispersion and mixing characteristics, (5) sunlight energy, (6) characteristics of the solids in the influent, and (7) the amount of essential microbial nutrients present. Basin efficiency, measured as the degree of stabilization of the incoming wastewater, is dependent on both biological process kinetics and basin hydraulic characteristics.

Three mechanisms affect the removal of organic compounds in aeration basins. These mechanisms are: biodegradation, adsorption on to the sludge, and air emissions.[18] Because these three mechanisms compete against each other, factors affecting the biodegradation and adsorption mechanisms will have an effect on air emissions. The greater the biomass concentration in the basin, the greater the removal of organic compounds will be by both biodegradation and adsorption mechanisms. The biodegradability of a compound will also affect the removal by biodegradation; as the biodegradability of the compound increases, so does the rate of biodegradation. Also, because the microorganisms prefer some compounds more than others, the biodegradation process is selective and depends on the compound matrix. Octanol-water partition coefficients are often used to indicate the affinity of a compound for the organic or aqueous phase. The relative magnitude of this coefficient provides some indication of organic compound removal by the adsorption mechanism.

Typically, aeration basins are equipped with aerators to introduce oxygen into the wastewater.[19] The biomass uses this oxygen in the process of biodegrading the organic compounds. However, aeration of wastewater also affects air emissions. Increased liquid surface area is exposed to ambient air, and, due to the turbulence caused by the aerators, the liquid and gas phase resistances to mass transfer are reduced. Convective mass transfer in both phases is increased. This transfer mechanism significantly increases air emissions compared to quiescent, flow-through type tanks like clarifiers. However, many of the factors which affect emissions from flow through tanks, like clarifiers, also affect emissions from aeration basins. The concentration and physical properties of the organics have a similar effect on emissions. As the volatility and diffusivities in water and air of the organic constituents increase, air emissions also tend to increase.

Other factors affecting emissions from aeration basins include wind speed and basin design characteristics. Figure 3-11 presents a typical aerated biological treatment basin. Increases in wind speed increase convective mass transfer from the wastewater in the basin and, therefore, increase air emissions. However, emissions from aeration basins are not as sensitive to wind speed effects compared to quiescent basins. Basin design characteristics which affect emissions include: the quiescent and turbulent surface areas, the depth of the basin, the design of the aerators, and the hydraulic retention time of the basin. As the turbulent surface area of the basin increases, air emissions will also tend to increase due to increased convective mass transfer of the organic compounds. The depth of the basin affects mass transfer in the liquid phase. Convective mass transfer in the liquid phase increases as the basin becomes more shallow, and, therefore, air emissions also tend to increase. Because the aerators generate the turbulence that increases the rate of mass transfer in the liquid and gas phases, the design of these aerators has a significant effect on emissions. The degree of turbulence these aerators impart to the wastewater is a function of the power output to the impellers, the impeller speed, and the impeller diameter. Increases in these design parameters result in additional turbulence of the wastewater and, for this reason, tend to increase air emissions. The final

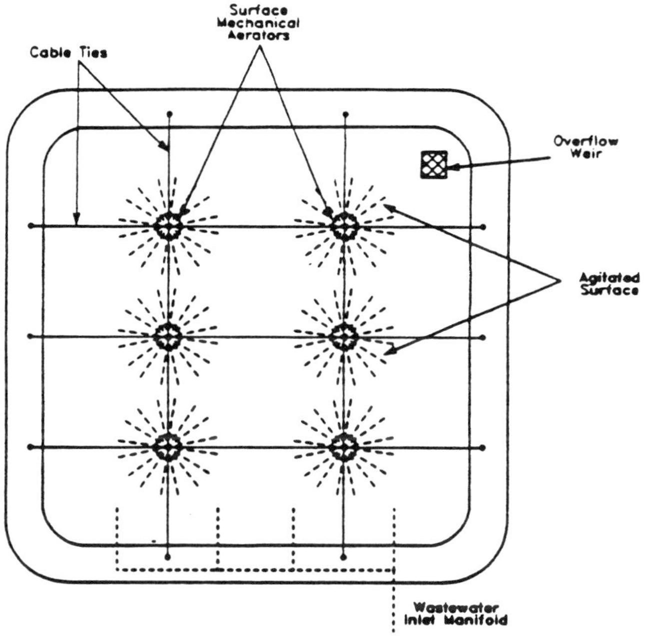

Design Parameter	Typical Design
Effective Diameter (m)	150
Surface Area (m^2)	17,650
Water depth (m)	2.0
Retention time (days)	6.5

Figure 3-11. Typical aerated biological treatment basin.

design parameter affecting emissions is the volume of the basin. As the volume increases, so does the hydraulic retention time. Increases in the basin volume provide additional time for removal by all three mechanisms: biodegradation, adsorption, and air emissions. Therefore, the magnitude of the increase in air emissions due to the additional retention time depends on the relative removal rates by the other two mechanisms. Development of biological treatment basin emission factors (aerated and nonaerated) is presented in Appendix A.

3.2.12 Treatment Tanks

Flocculation tanks and pH adjustment tanks are typically used for treatment of wastewater after and before biological treatment, respectively. In flocculation tanks, flocculating agents are added to the wastewater to promote formation of large particle masses from the fine solids formed during biological treatment. These large particles will then precipitate out of the wastewater in the clarifier which typically follows. Tanks designed for pH adjustment typically precede the biological treatment step. In these tanks, the wastewater pH is adjusted, using acidic or alkaline additives, to prevent shocking of the biological system downstream.

3.2.12.1 VOC Emission Mechanism. Emissions occur from treatment tanks by both diffusive and convective mechanisms.[20] As wastewater flows slowly through the tank, organic compounds diffuse through the water to the liquid surface. These organic compounds volatilize into the ambient air above the liquid and vapor phases. Since the organic vapors above the liquid are in contact with the ambient air, these organic vapors can be swept into the air by wind blowing across the treatment tank surface.

3.2.12.2 Factors Affecting Emissions from a Treatment Tank. The factors affecting emissions from a treatment tank are similar to the factors affecting emissions from other well mixed, flow-through impoundments.[21] These factors are: the wastewater characteristics, wind speed, and design characteristics of the treatment tank. Increases in the wastewater temperature, organic compound concentration, and compound physical properties such as volatility and diffusivity in water increase emission rates to the air. Likewise, increases in wind speed improve convective mass transfer, thereby increasing air emissions. The design characteristics of the treatment tanks that affect emission rates are the liquid surface area, the fetch to depth ratio, and the hydraulic retention time. Increases in the magnitude of these design parameters increase emission rates to the air. Development of treatment tank emission factors is presented in Appendix A.

3.2.13 Surface Impoundments

Surface impoundments are typically used for evaporation, polishing, equalization, storage prior to further treatment or disposal, equalization, leachate collection, and as emergency surge basins. They could be quiescent or mechanically agitated.

3.2.13.1 VOC Emission Mechanism. Emissions occur from surface impoundments by both diffusive and convective mechanisms. As wastewater flows slowly through the tank, organic compounds diffuse through the water to the liquid surface. These organic compounds volatilize into the ambient air above the liquid and vapor phases. Since the organic vapors above the liquid are in contact with the ambient air, these organic vapors can be swept into the air by wind blowing across the surface of the impoundment.

3.2.13.2 Factors Affecting Emissions from a Surface Impoundment. The factors affecting emissions from a surface impoundment are similar to the factors affecting emissions from equalization basins if it is quiescent and similar to factors affecting emissions from aeration basins if it is agitated. Emission factor development for a surface impoundment will vary depending on its purpose.

All characteristics of the impoundment should be reviewed to determine what type of collection or treatment system it best resembles. Once the surface impoundment has been characterized, refer to Appendix A for determination of emission factors.

3.3 EXAMPLE WASTE STREAM COLLECTION AND TREATMENT SYSTEM SCHEMATICS

Emission mechanisms and factors affecting emissions from wastewater collection and treatment units were presented in the previous section. The general scheme used to collect and treat process wastewater varies from facility to facility and depends on many factors. Some factors which affect the general scheme used at a particular facility are: the compounds contained in the wastewater streams leaving different process areas, the flow rate of these streams, the general equipment layout in the process areas, the terrain around the facility, and the age of the facility. Also organic compounds volatilize at different rates from different collection and treatment units. Therefore, the general scheme used to collect and treat wastewater at a facility has a significant impact on VOC emissions.

Due to the many factors that affect the general scheme used to collect and treat facility wastewater, it is not possible to develop example wastewater streams representing all possible scenarios. In lieu of this, three example waste stream collection and treatment schemes were developed to evaluate potential ranges in emissions from different facilities. The collection and treatment system schematics presented were chosen to represent a range of emission potentials and are not meant to characterize specific facilities in the industries covered by this document.

Schematics of the three example waste stream collection and treatment systems are shown in Figures 3-12, 3-13, and 3-14. Each figure shows the discharge of a process wastewater stream into the example waste stream collection system. In each case, the wastewater stream proceeds through the collection system components and is combined for treatment with other facility wastewater streams. Dimensions of these collection and treatment system units are presented in Table 3-2. These dimensions are based on typical designs for each unit and are within the dimension ranges discussed in Section 3-2.

For purposes of comparison, emissions are estimated for a wastewater stream with the same flow rate and organic concentration as it flows through the collection and treatment units in each example schematic. The wastewater stream is discharged from the process equipment at a flow rate of 300 liters per minute (lpm) and an overall organic concentration of 2,500 parts per million (ppm). Table 3-3 provides a description of the example wastewater stream. This example wastewater stream was designed to contain compounds that span the range of volatilities to demonstrate a range of emission potentials. Emissions are estimated from the collection and treatment units in each example waste stream system using techniques presented in Appendix A. These emission estimates are presented in Tables 3-4, 3-5, and 3-6 for Example Waste Stream Schematics I, II, and III, respectively. In each table, emissions are presented as a fraction of the mass of organic compounds entering the

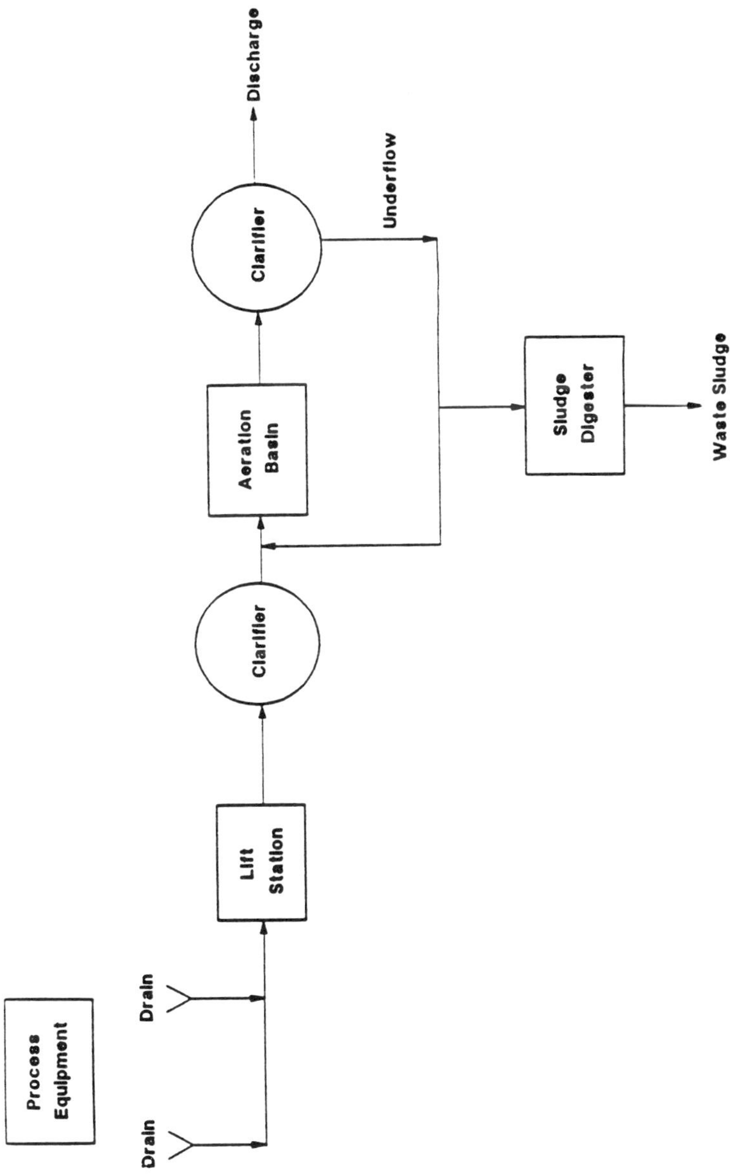

Figure 3-12. Example Waste Stream Schematic I.

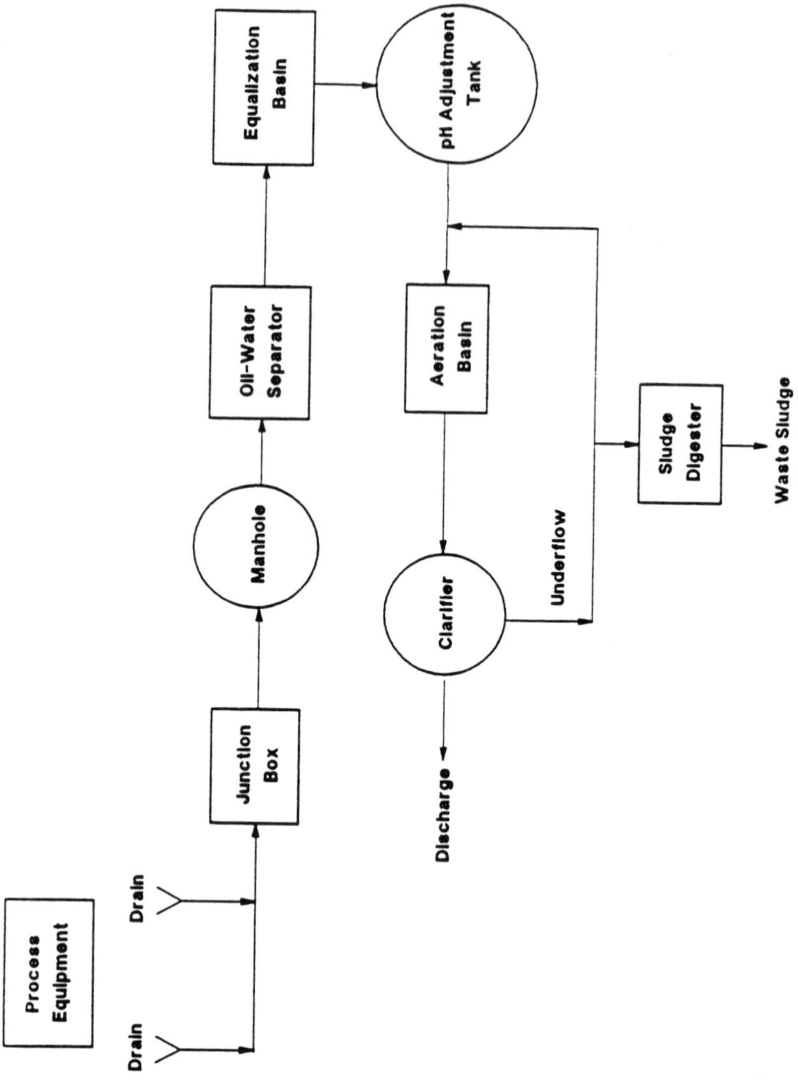

Figure 3-13. Example Waste Stream Schematic II.

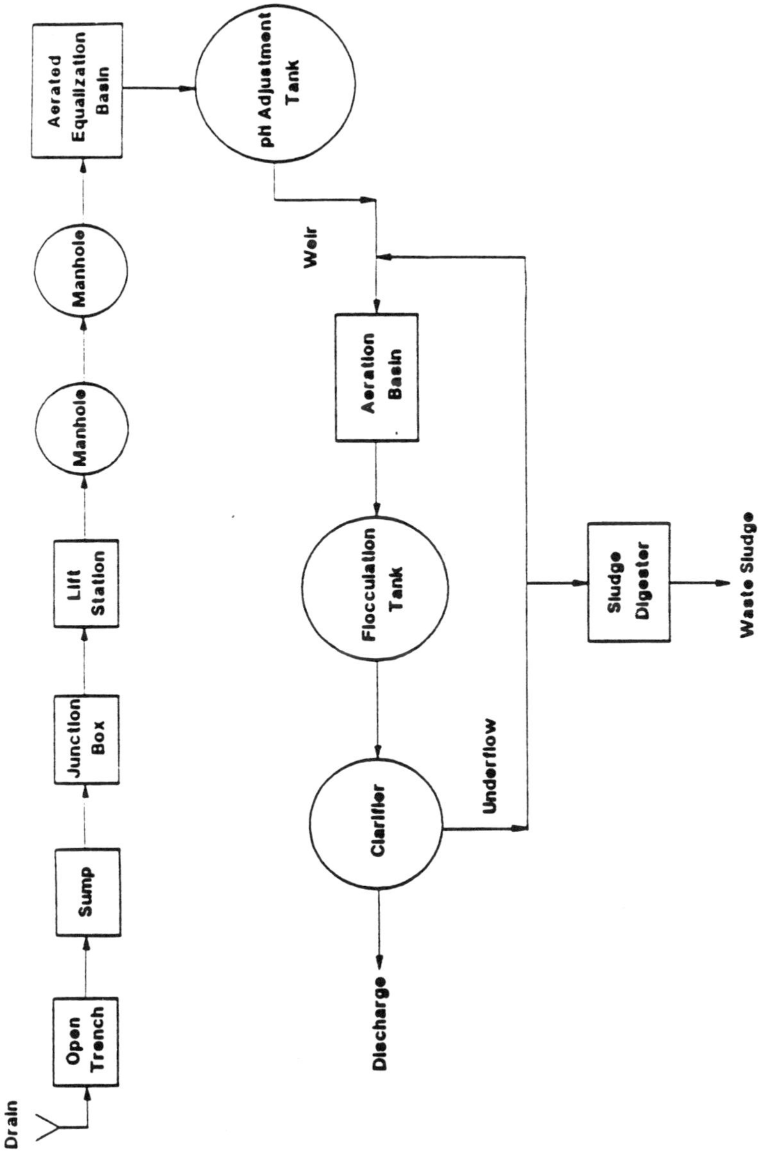

Figure 3-14. Example Waste Stream Schematic III.

TABLE 3-2. DIMENSIONS FOR EXAMPLE WASTE STREAM COLLECTION AND TREATMENT UNITS

Component	Design Parameter	Typical Dimensions
Drain	riser height (m)	0.6
	riser diameter (m)	0.2
	process drain pipe diameter (m)	0.1
	effective diameter of riser (m)	0.1
	riser cap thickness (cm)	0.6
	sewer diameter (m)	0.9
Manhole	diameter (m)	1.2
	height (m)	1.2
	cover diameter (m)	0.6
	diameter of holes in cover (cm)	2.5
	cover thickness (cm)	0.6
	sewer diameter (m)	0.9
Junction Box	effective diameter (m)	0.9
	grade height (m)	1.5
	water depth (m)	0.9
	surface area (m^2)	0.7
Lift Station	effective diameter (m)	1.5
	width, (m)	1.8
	grade height (m)	2.1
	water depth (m)	1.5
	surface area (m^2)	1.8
Trench	length (m)	15.2
	water depth (m)	0.6
	depth (m)	0.8
	width (m)	0.6
Weir	height (m)	1.8
Oil/Water Separator	length (m)	13.7
	width (m)	7.6
	retention time (hr)	0.8
Clarifier	diameter (m)	18.3
	depth (m)	3.5
	retention time (hr)	4.0

TABLE 3-2. (Continued)

Component	Design Parameter	Typical Dimensions
Sump	effective diameter (m)	1.5
	water depth (m)	1.5
	surface area (m^2)	1.8
Equalization Basin	effective diameter (m)	109
	water depth (m)	2.9
	surface area (m^2)	9,290
	retention time (days)	5
Aeration Basin	effective diameter (m)	150
	water depth (m)	2.0
	surface area (m^2)	17,652
	retention time (days)	6.5
Treatment Tank	effective diameter (m)	11
	water depth (m)	4.9
	surface area (m^2)	93
	retention time (hr)	2

TABLE 3-3. EXAMPLE WASTEWATER STREAM

Wastewater Stream Content: Water = 99.75%
Total Organics = 0.25% (2,500 ppm)

Wastewater Flow: 300 lpm

Organic Compound	Waste Stream Concentration (ppm)	Volatility Category	Henry's Law (25°C) (atm-m^3/gmole)
Butadiene	500	High	1.42×10^{-1}
Toluene	500	Medium	6.68×10^{-3}
Naphthalene	500	Medium	1.18×10^{-3}
Butanol	500	Low	8.90×10^{-6}
Phenol	500	Low	4.54×10^{-7}
Total:	2,500		

TABLE 3-4. EMISSION ESTIMATE FOR EXAMPLE WASTE STREAM SCHEMATIC I

Component[a]	Fraction Emitted[b,c]	Cumulative Fraction Emitted[b,c]
Drain	0.13	0.13
Drain	0.13	0.19
Lift Station	0.17	0.29
Clarifier	0.017	0.30
Aerated Biological Treatment	0.12	0.36
Clarifier	0.017	0.36

[a]Collection and treatment system components defined in Table 3-2.
[b]Based on wastewater stream described in Table 3-3.
[c]Calculations presented in Appendix A.

TABLE 3-5. EMISSION ESTIMATE FOR EXAMPLE WASTE STREAM SCHEMATIC II

Component[a]	Fraction Emitted[b,c]	Cumulative Fraction Emitted[b,c]
Drain	0.13	0.13
Drain	0.13	0.19
Junction Box	0.057	0.23
Manhole	0.032	0.23
Oil/Water Separator	0.31	0.35
Non-Aerated Equalization Basin	0.28	0.44
pH Adjustment	0.0062	0.44
Aerated Biological Treatment	0.12	0.45
Clarifier	0.017	0.45

[a]Collection and treatment system components defined in Table 3-2.
[b]Based on wastewater stream described in Table 3-3.
[c]Calculations presented in Appendix A.

TABLE 3-6. EMISSION ESTIMATE FOR EXAMPLE WASTE STREAM SCHEMATIC III

Component[a]	Fraction Emitted[b,c]	Cumulative Fraction Emitted[b,c]
Drain	0.13	0.13
Open Trench	0.026	0.15
Sump	0.0035	0.15
Junction Box	0.057	0.19
Lift Station	0.17	0.30
Manhole	0.032	0.30
Manhole	0.032	0.31
Aerated Equalization Basin	0.73	0.73
pH Adjustment Tank	0.0062	0.74
Weir	0.26	0.80
Aerated Biological Treatment	0.12	0.81
Flocculation Tank	0.0062	0.81
Clarifier	0.017	0.81

[a]Collection and treatment system components defined in Table 3-2.
[b]Based on wastewater stream described in Table 3-3.
[c]Calculations presented in Appendix A.

collection and treatment system. Tables 3-4, 3-5, and 3-6 show the cumulative fraction of these organic compounds that is emitted at each stage in the collection and treatment systems for each waste stream system. The cumulative fraction emitted from Example Waste Stream Schematics I, II, and III are 0.36, 0.45, and 0.81, respectively.

Table 3-7 presents the estimates of annual VOC emissions from each of these example schematics for various flow rates. From an evaluation of the results presented in this table, it can be seen that overall VOC emissions are lowest for Example Waste Stream Schematic I and highest for Example Waste Stream Schematic III. Because the wastewater stream entering each example stream is identical, the difference in overall VOC emissions is due to the presence of different collection and treatment system components in the three example schematics. Schematic I has the least number of collection and treatment system components; and, therefore, has the lowest overall VOC emission factor. Schematic III has the greatest number of collection and treatment system components and; therefore, the highest overall VOC emission factor. Although the number of collection and treatment system components may indicate the emission potential, the characteristics of these components also affects the emission potential. Sources which are quiescent or have small surface areas would have lower overall emission factors than sources which are aerated or turbulent. Clarifiers, treatment tanks, sumps, trenches, and junction boxes are typically lower emitters than aerated equalization basins, biological treatment, weirs, lift stations, and drains.

TABLE 3-7. SUMMARY OF THE ESTIMATED ANNUAL VOC EMISSIONS FROM EACH OF THE EXAMPLE WASTE STREAM SCHEMATICS

		VOC Emissions (Mg/yr)[a]				
Example Schematics	Emission Factor[b] (fe)	Wastewater Flow Rate (lpm)				
		40	150	300	455	760
Schematic I	0.36	18.7	70	140	213	355
Schematic II	0.45	23.9	90	180	272	445
Schematic III	0.81	42.4	159	318	482	805

[a]The assumed wastewater volatile organic concentration is 2,500 ppm. (See Table 3-3 for the characteristics of the example waste stream.)
[b]Overall cumulative fraction emitted for each Example Waste Stream Schematic. (See Tables A-33 through A-35 in Appendix A.)

3.4 REFERENCES

1. Office of Air Quality Planning and Standards, U. S. Environmental Protection Agency. Research Triangle Park, North Carolina. VOC (Volatile Organic Compound) Emissions from Petroleum Refinery Wastewater Systems - Background Information for Proposed Standards, Draft EIS. EPA-450/3-85-001a (NTIS PB87-190335), February 1985.

2. Reference 1.

3. Reference 1.

4. Reference 1.

5. Office of Air Quality, Planning and Standards. U. S. Environmental Protection Agency. Hazardous Waste Treatment, Storage, and Disposal Facilities (TSDF) - Air Emission Models, Draft Report. April, 1989.

6. Reference 5.

7. Reference 5.

8. Lyman, Warren, Ph.D., Reehl, William F., and Rosenblatt, David H., Ph.D. Handbook of Chemical Property Estimation Methods. McGraw-Hill Book Company, New York, New York, 1982. pp. 15-9 to 15-31.

9. Reference 8.

10. Berglund, R. L., and Whipple, G. M. "Predictive Modeling of Organic Emissions", Chemical Engineering Progress. Union Carbide Corporation, South Charleston, West Virginia, pp. 46 - 54.

11. Reference 10.

12. Liang, S. F. Hydrocarbon Losses by Atmospheric Evaporation from Open Separators, Technical Progress Report, BRC-CORP24-73-F. (Shell Models).

13. Reference 5.

14. Reference 5.

15. Metcalf & Eddy, Inc. Wastewater Engineering: Treatment/Disposal/Reuse. McGraw-Hill Book Company, New York, New York, 1979. pp. 201-222.

16. Reference 5.

17. Reference 5.

18. Reference 5.

19. Reference 15.

20. Reference 5.

21. Reference 5.

4. VOC Emission Control Equipment

As discussed in Chapter 3, VOC emissions during collection and treatment of industrial wastewater can be significant, and measures to control these emissions need to be considered. This chapter describes control measures that can be applied to reduce these VOC emissions. Three control strategies are discussed in this chapter. The first control strategy is waste minimization through process modifications, modification of operating practices, preventive maintenance, recycling, or segregation of waste streams. The second control strategy is to reduce the organic content of the wastewater through treatment before the stream contacts ambient air. The third strategy is to control emissions from collection and treatment system components until the organic compounds are either recovered or destroyed. Although the third strategy is feasible in some cases, the more universally applicable treatment technology is to reduce the quantity of waste generated or reduce the organic content of the wastewater at the point of generation.

One type of treatment technology available and currently in use at many facilities is steam stripping. Because steam stripping removes the organic compounds most likely to be emitted downstream (most volatile compounds), it is an effective technique for reducing VOC emissions from wastewater. However, in some applications another organic removal technique may be better suited. In other cases it may be more reasonable to control emissions up through removal or destruction of the organic compounds. The purpose of this section is to present and discuss some of the various emission control strategies. A general discussion of the application of waste minimization to control VOC emissions from industrial wastewaters is presented in Section 4.1. Section 4.2 presents a discussion of organic compound removal technologies. Section 4.3 presents VOC emission suppression controls from collection and treatment system components. Add-on control devices are discussed in Section 4.4.

4.1 WASTE MINIMIZATION

Waste minimization is a general term which includes both source reduction and recycling. Source reduction refers to reduction or elimination of the generation of a specific waste at the source. This may be accomplished through process or equipment modifications, stream segregation, or changes in work practices. Recycling includes recovery and/or reuse of potential waste streams. Waste minimization must be implemented on a process-specific basis. However, implementation of an aggressive waste minimization program can be an effective method of reducing emissions of VOC from industrial wastewaters.

Although many of the specific techniques which can be applied to minimize waste generation are specific to one application, the implementation of any waste minimization program should follow the guidelines presented below. By following these guidelines, the most effective steps can be identified and implemented.

4.1.1 Gather Baseline Data

The first step in any waste minimization program should be to identify and characterize the individual waste streams. This should include flow rate, composition, pH, and solids content of the wastewater streams. Although some of this data might need to be gathered through a sampling program, some of it may be available from hazardous waste manifests, Superfund Amendments and Reauthorization Act (SARA) Title III Section 313 release reporting calculations, permits, monitoring reports, product and raw material specifications, and other internal records.

4.1.2 Identify and Rank Sources for Reduction

Using the baseline data gathered, a cost allocation system should be developed to assess treatment and disposal costs to individual waste streams. Future treatment and disposal costs should be considered in this evaluation, as should potential liabilities associated with the waste handling and subsequent treatment and disposal. Once the waste streams have been ranked and prioritized, methods for controlling these streams can be considered.

4.1.3 Implementation of Reduction/Recycling

In selecting the appropriate method for reducing or eliminating a wastewater stream, a variety of sources of information can be utilized. EPA's Pollution Prevention Information Clearinghouse (PPIC), supported by the EPA's Pollution Prevention Office contains information on case studies and reports on pollution prevention. PPIC can be accessed by telephone hotline. Other valuable sources of information are State assistance programs, vendors, and consultants.

As waste minimization steps are implemented, it is important that good record keeping be continued to document which steps were effective and which ones failed. This is especially important since future regulations may require percentage reductions in wastes generated. Although some wastewater streams will still be generated, an effective waste minimization program may allow more cost-effective handling of these streams.

4.2 ORGANIC COMPOUND REMOVAL

4.2.1 Steam Stripping

Steam stripping is a proven technology which involves the fractional distillation of wastewater to remove organic compounds. The basic operating principle of steam stripping is the direct contact of steam with wastewater. This contact provides heat for vaporization of the more volatile organic compounds. The overhead vapor containing water and organics is condensed and separated (usually in a decanter) to recover the organics. These recovered organics are usually either recycled or incinerated in an on-site combustion device.

In principle, a multistage steam stripper system can be designed to achieve almost any level of organic compound removal. In practice, the achievable VOC emission reductions and associated control costs are highly dependent on wastewater characteristics such as flow, organic concentration and composition, and the design of the collection and treatment systems.

As previously discussed, based on industry responses to Clean Air Act Section 114 information requests, 20 percent of the reported wastewater streams account for 65 percent of the organics by mass.[1] Therefore, it may be possible to achieve significant VOC emission reduction by controlling a relatively small number of individual wastewater streams containing organic compounds. In many cases, it may be possible to combine two or more of these streams for treatment by the same steam stripper by hard piping these streams from the point of generation to the steam stripper. As streams are combined, the cost of control increases, however, cost per stream decreases. In addition, the emission reduction achieved by controlling the combined streams increases. This issue is discussed further in the presentation of steam stripper costs in Chapter 6.

4.2.1.1 Steam Stripper Process Description. Steam stripper systems may be operated in batch or continuous mode. Batch steam stripping is more prevalent when the wastewater feed is generated by batch processes.[2] Batch strippers may also be used if the wastewater contains relatively high concentrations of solids, resins, or tars. Usually, batch steam strippers provide a single equilibrium stage of separation. Therefore, the removal efficiency is essentially determined by the equilibrium coefficients of the pollutants and the fraction of the initial charge distilled overhead.

Wastewater is charged to the receiver, or pot, and brought to the boiling temperature of the mixture. Heat is provided by direct injection of steam or by an external heat exchanger normally referred to as a reboiler. The overhead vapors are condensed and recovered. The solids, tars, resins, and other residue remaining in the pot are normally disposed. By varying the heat input and fraction of the initial charge boiled overhead, the same batch stripper can be used to treat wastewater mixtures with widely varying characteristics.

In contrast to batch strippers, continuous steam strippers are normally designed to treat wastewater streams with relatively consistent characteristics.[3] Design of the continuous stripper system is normally based on the flow rate and composition of a specific wastewater feed stream or combination of streams. Multistage, continuous strippers normally operate at greater organic compound removal efficiencies than batch strippers. Continuous systems may also offer other advantages (over batch stripping) for applications involving wastewater streams with relatively high flows and consistent concentrations. These advantages include more consistent effluent quality, more automated operation, and lower annual operating costs.

As discussed in the introduction for Section 4.2.1, it may be possible to achieve significant emission reduction by controlling a relatively small number of wastewater streams containing organic compounds. Wastewater streams that are continuously discharged from process equipment are usually relatively consistent in composition. Such wastewater streams would be treated with a continuous steam stripper system. However, batch wastewater streams can also be controlled by continuous steam strippers by incorporating a feed tank with adequate residence time to provide a relatively consistent outlet composition. For these reasons, the remaining discussion focuses on continuous steam stripping.

A continuous steam stripper system is shown in Figure 4-1. The steam stripper presented in this figure is a generic stripper. For specific cases, modifications to the design may provide more effective control. Alternate feed locations and multiple feed locations are sometimes used. The steam stripper can also be operated under a vacuum. In addition, steam strippers may include a reflux stream where the bottoms stream flows into a reboiler to vaporize and return to the column a portion of the bottoms stream. At least one steam stripper studied, operated at an altered pH to change the equilibrium of a reaction of a less volatile compound. The low volatility compound through this reaction was removed efficiently.

In designing the appropriate steam stripper, these refinements should be considered. However, the generic steam stripper presented in Figure 4-1 is a generally applicable control which can be applied to control wastewater streams with significant concentrations of organic compounds. The purpose, design, and operation of each of the components of this system (a wastewater feed tank, feed/bottoms heat exchanger, condenser system, and necessary pumps) are discussed separately in this section.

The first component in the steam stripper system is the controlled sewer system or hard piping from the point of wastewater generation to the feed tank. This is necessary to control emissions prior to steam stripping. The feed tank collects and conditions the wastewater feed to the steam stripper. The tank is covered and vented to an on-site combustion device. Sufficient residence time is provided to ensure that solids, oils, and grease that could foul the heat exchanger and stripping column are separated from the wastewater in the feed tank. Residence time in the feed tank also ensures that wastewater is fed to the steam stripper at a relatively consistent flow rate and composition.

After the wastewater is collected and conditioned, it is pumped from the feed tank through the feed/bottoms heat exchanger and into the steam stripping column. Steam is usually directly injected into the stripper at the bottom of the column. The wastewater feed is usually introduced into the stripper at the top of the column as shown in Figure 4-1. As the wastewater flows down through the column it contacts this steam which is flowing countercurrently up the column. Organic compounds in the wastewater absorb heat from the steam and volatilize into the vapor stream. These constituents flow out the top of the column with any uncondensed steam.

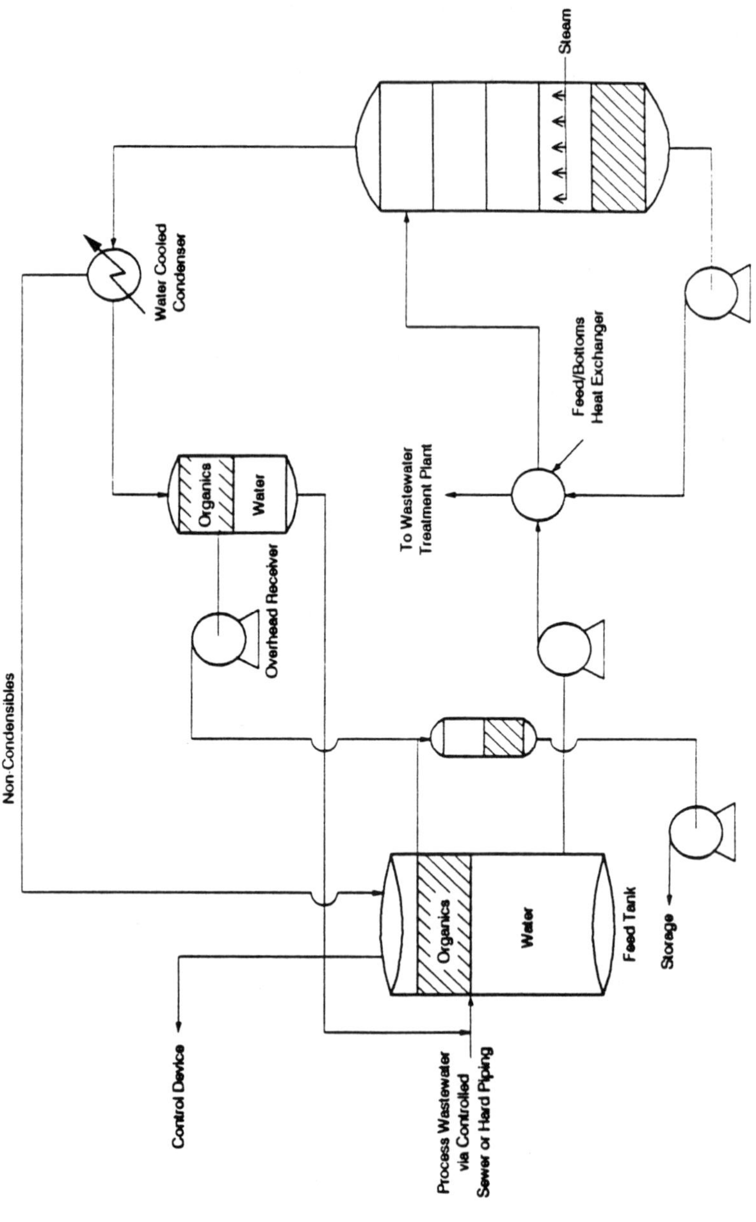

Figure 4-1. Continuous Steam Stripper System

A condenser system is used to recover the organic and water vapors present in the overheads stream. The condensed overheads stream is fed to a decanter where the organic and water phases are gravity separated. The organic phase is usually either pumped to storage and then recycled to the process or burned as fuel in a combustion device. The water phase is returned to the feed tank and recycled through the steam stripper. Any non-condensable gases and highly volatile organic compounds not recovered by the condenser system are routed to an on-site control device such as a carbon adsorber, boiler, or an incinerator. (In Figure 1, the noncondensables are vented to the feed storage tank, which is routed to a control device.) If the primary condenser is not sufficient for condensing a large portion of the organics, it may be necessary to install a secondary condenser with brine or a refrigerant.

The wastewater effluent leaving the bottom of the steam stripper is pumped through the feed/bottoms heat exchanger. This serves to heat the feed stream and cool the bottoms prior to discharge. This exchange of waste heat also improves the economy of the steam stripper system. After passing through the heat exchanger, the bottoms stream is usually either routed to an on-site wastewater treatment plant as shown in Figure 4-1 and discharged to an National Pollutant Discharge Elimination System (NPDES) permitted outfall, or is sent to a POTW.

4.2.1.2 Steam Stripper Design and Operation Information on the design and operation of steam stripper systems is available from studies conducted by the EPA. During the Industrial Wastewater Project, EPA obtained information on approximately 15 steam strippers from facility responses to Clean Air Act Section 114 information requests.[4] In addition, during this project, information was gathered on site visits to nine chemical manufacturing facilities operating steam strippers to remove organic compounds from wastewater.[5-13] The EPA also gathered data on steam stripper operation as a part of the Hazardous Waste Treatment, Storage, and Disposal Facilities (TSDF) Project. During this project, data were gathered on three steam strippers through field testing efforts.[14-16]

The EPA gathered data on steam strippers during the development of effluent guidelines for the OCPSF, Pesticide, and Pharmaceutical Manufacturing industries. In response to Clean Water Act Section 308 information requests, 63 OCPSF facilities reported using steam strippers as an in-plant control for process wastewater.[17] A total of 108 steam strippers are reportedly operated by these facilities. In addition to these information requests, data on steam strippers in operation at four OCPSF facilities were obtained through field testing efforts.[18-21] Less data are available for steam strippers in use at Pesticide and Pharmaceutical Manufacturing industry facilities. Steam strippers are reportedly used at eight pesticide facilities[22] and eight pharmaceutical facilities.[23]

The steam stripper systems discussed above are used to treat a variety of wastewater streams. These wastewater streams vary in flow rate and composition. In addition, some streams contain relatively high levels of suspended solids.

Although the wastewater characteristics vary, the basic steam stripper system shown in Figure 4-1 can be designed and operated to achieve high organic compound removal efficiencies for most streams. Table 4-1 presents the design and operating conditions for a steam stripper with an assumed wastewater feed rate of 300 liters per minute (lpm). The design was developed using Advanced System for Process Engineering (ASPEN).[24] ASPEN is a computer software program intended for the rigorous design of distillation columns. The major design parameters in the ASPEN steam stripper model are based on field experience and published information. These include a pressure drop of 0.41 kPa per meter of packing, 60 percent of calculated flooding for the packing, and 2.54 cm stainless steel saddles (random dumped packing). In addition, the following engineering assumptions were made:

- Operating pressure of one atmosphere;
- Isothermal column operation;
- Constant molal overflow (i.e., one mole of aqueous phase vaporized for each mole of steam condensed); and
- Linear equilibrium and operating equations (i.e., Henry's Law is valid for each organic compound at the concentrations encountered in the stripping column).

Because the ASPEN Model is a packed tower, some engineering assumptions were made to approximate a tray column design. ASPEN calculates the number of transfer units (NTU). A conversion factor of 0.53 theoretical trays per transfer unit was used by ASPEN to determine the number of theoretical trays. A tray efficiency of 75 percent was assumed to estimate the actual number of stages for the column. A tray spacing of 0.46 meters was assumed to estimate

TABLE 4-1. DESIGN AND OPERATING BASIS FOR THE STEAM STRIPPING SYSTEM

1. Wastewater Stream Content: water = 99.75%
total organics = 0.25% (2,500 ppm)

2. Wastewater Stream Organic Composition:

Volatility Category	Organic Compound	Organic Compound Henry's Law Value (atm - m^3/gmol) at 25°C	Waste Stream Content (ppm)	% Removal in Stripper[a]
High	Butadiene	1.42×10^{-1}	500	100
Medium	Toluene	6.68×10^{-3}	500	100
Medium	Naphthalene	1.18×10^{-3}	500	99.9
Low	Butanol	8.90×10^{-6}	500	30
Low	Phenol	4.54×10^{-7}	500	2.2

3. Wastewater Flow: 300 l/min

4. Stripper Operating Period: 24 hr/day x 300 day/yr = 7,200 hr/yr

5. Wastewater Storage: Wastewater feed collection tank with 48 hr retention time

6. Steam Stripping Column:
Configuration: countercurrent flow, sieve tray column
Steam Flow Rate: 0.06 kg of steam /liter of waste feed
Wastewater Feed Temperature: 35°C
Column Diameter: 0.76 m
Active Column Height: 14.2 m
Total Column Height: 18.3 m
Liquid Loading: 39,900 l/hr/m^2

7. Condenser:
Configuration: Water-cooled
Primary Condenser Outlet Vapor Temperature: 50°C

8. Overhead Control: Vent to existing on-site combustion or other control device

9. Bottoms Control: Feed to existing on-site wastewater treatment facility or POTW

[a]Removal efficiency was estimated using ASPEN.[24] Benzene was the chosen design compound.

the active column height. To approximate the total column height, 4.1 meters of nonactive entrance and exit column was assumed.

Using ASPEN, the most cost-effective feasible design for controlling the model waste streams presented in Chapter 3 was developed. The associated costs of this steam stripper are presented in Chapter 6.

For purposes of evaluation, the steam stripper system can be separated into the following three functional sections: collection and conditioning of the wastewater, steam stripping of the wastewater, and recovery of the steam stripped organics. Each of these sections is discussed in detail below as they pertain to well-designed and operated systems.

4.2.1.2.1 Collection and conditioning of the wastewater. In controlling VOC emissions from wastewaters using a steam stripper, the first necessary step is to control emissions from the point of wastewater generation to the steam stripper feed tank. This can be accomplished by hardpiping the wastewater to the tank, or by controlling the collection system components as discussed in Section 4.3.1.. Next, the steam stripper feed tank should be controlled by venting to an on-site incinerator or other control device. The stripper feed tank is used to collect and condition wastewater feeding the steam stripper. The feed tank is normally sized to provide a desired hydraulic retention time of 0.5 to 40 hours for the wastewater feed stream.[25,26] However, for controlling individual batch streams, additional storage capacity may be required. The desired retention time depends primarily on two factors: (1) the variability in wastewater flow from the source, and (2) the amount of conditioning required prior to steam stripping. Additional retention time is required to provide surge capacity for wastewater streams with highly variable flow rates. As the wastewater flow rate from the process varies, the level in the feed tank is also allowed to vary to provide a relatively constant feed rate to the stripper. If the feed tank is adequately designed, a continuous steam stripper may even be used to treat wastewater generated by some batch processes. In these cases, the feed tank serves as a buffer between the batch process and the continuous steam stripper. During periods of no wastewater flow from the batch process, stored wastewater is fed to the stripper at a relatively constant rate, and the feed tank level is allowed to decrease. The tank design is based on an assumed retention time of 48 hours. This allows sufficient residence time for multiple batch or continuous streams to be combined to provide a constant wastewater composition to the steam stripper.

As mentioned earlier, the desired retention time in the feed tank also depends on the degree of conditioning required for the stripper feed stream. Water and organic phases are often present in the stripper feed tank. The feed tank provides the retention time necessary for these phases to separate due to density differences. Oils and tars which may also be present normally partition from the water into the organic phase. This organic phase is normally either recycled to the process for recovery of the organic compounds or disposed of by incineration. The water phase is fed to the stripper to remove the soluble organics. Solids are also separated in the stripper feed tank. The separation efficiency depends on the density of the solids dissolved in the process wastewater. Some of the lighter solids may remain dissolved in the organic or water phases present in the feed tank. However, the heavier solids will settle to the bottom of the tank. These solids are normally removed from the feed tank periodically and usually landfilled.

If excessive levels of oils, grease, or solids are present, it may be necessary to install additional equipment to assist the feed tank in conditioning the stripper feed stream. In some stripper system designs, the wastewater first flows through an oil/water decanter before proceeding to the stripper feed tank.[27] The decanter provides an additional separation stage for removal of oils and grease that could cause fouling and plugging in the stripper column. However, decanters or oil/water separators used to pretreat wastewater should be covered and controlled. High dissolved solids levels may also cause fouling and plugging problems in the column. In these cases, it may be more efficient to separate the solids prior to the feed tank using filtration equipment. However, the column internals can sometimes be modified to accommodate relatively high solids concentrations. High solids levels may be more of a problem for system components other than the column. These problems include plugging of the feed/bottoms heat exchanger and cause excessive wear on the stripper feed and bottoms pumps.

4.2.1.2.2 Steam stripping of the wastewater. Steam stripper columns are designed and operated to ensure that sufficient contact occurs between the wastewater containing the organic compounds and the steam which provides the heat energy. Three main factors affect the degree of contact that occurs in the column. These are: (1) the dimensions of the column (height and diameter), (2) the contacting media used in the column (trays or packing), and

(3) operating parameters such as the steam to feed ratio, column temperatures, and pH of the wastewater. The organic removal performance of the steam stripper depends on the degree of contact between steam and wastewater and is, therefore, a function of the design and operating values selected for these parameters.

The diameter of the column determines the cross-sectional area available for liquid and vapor flow through the column. This cross-sectional area must be sufficient to prevent flooding due to excessive liquid loading or liquid entrainment due to excessive vapor velocity.[28] The cross-sectional area also affects the liquid retention time in the column. Column diameters reported in Section 114 responses range from 0.3 m to 1.8 m, and the steam stripper presented in Table 4-1 has a diameter of 0.76 m.[29]

The contacting media used in the column plays a major role in determining the mass transfer efficiency. Generally, steam stripping columns are equipped with trays or packing to provide contact between the vapor and liquid phases. Trays are regularly spaced throughout the column and provide staged contact between the two phases; packing provides for continuous contact. There are advantages and disadvantages to the use of both trays and packing as the contacting media in steam stripping columns.[30] Trays are generally more effective for wastewater containing dispersed solids due to plugging and cleaning problems encountered with packing. In addition, tray towers can operate efficiently over a wider range of liquid flow rates than packed towers. Packed towers are often more cost effective to install and operate when treating wastewater streams that are highly corrosive. Also, the pressure drop through packed towers may be less than the pressure drop through tray towers designed for equivalent wastewater loadings. However, packed towers are seldom designed with diameters in excess of four feet and column heights may be more limited (compared to tray towers) due to crushing of the packing located near the bottom of the column.

The column height is determined by the number of theoretical stages required to achieve the desired VOC removal. The number of theoretical stages is a function of the equilibrium coefficient of the pollutants and the efficiency of mass transfer in the column.[31] (As mentioned in Section 4.2.1.2, the number of actual stages was approximated assuming a tray efficiency of 75 percent.) The steam stripper is a 18.3 m sieve tray countercurrent flow column, with 31 sieve trays.

In addition to the design factors discussed above, several operating parameters affect the organic compound removal performance of steam strippers. An increase in the steam to feed ratio will increase the ratio of the vapor to liquid flow through the column. This increases the stripping of organics into the vapor phase. Because additional heat is provided when the steam rate is increased, additional water is also volatilized. Therefore, an increase in the steam to feed ratio is also normally accompanied by an increase in the steam rate flowing out of the column in the overheads stream.

Steam to feed ratios generally range from 0.01 kg/kg to 1.0 kg/kg. A steam to feed ratio of 0.06 was assumed for the model steam stripper. The steam to feed ratio required for high removal efficiency depends strongly on the temperature of the wastewater feed to the column. If the feed temperature is lower than the operating temperature at the top of the column, part of the steam is required to heat the feed. Good column design requires that sufficient steam flow is available to heat the feed and volatilize the organic constituents. Steam in excess of this sufficient flow rate helps to carry the organic compounds out of the top of the column with the overheads stream.

Column operating temperature and wastewater pH also affect organic compound removal performance.[32] Temperature affects the solubility and equilibrium coefficients of the organic compounds. Usually stripping columns are operated at pressures slightly greater than atmospheric. Therefore, in relation with the column pressure, column temperatures usually operate slightly greater than the normal boiling point of water. Wastewater pH is often controlled by adding caustic to the feed to ensure that these types of organic compounds remain in a steam strippable form.[33]

4.2.1.2.3 Recovery of the steam stripped organics. The conceptual design shown in Figure 4-1 employs a one-stage condenser system. Any condensables not removed from the vapor stream in the primary condenser are vented to a control device. The secondary condenser is normally chilled (brine water or other coolant) to ensure condensation of the more volatile organic compounds. In Table 4-1, the steam stripper outlet vapor condenser temperature for the primary condenser is 50°C. The condensed steam and organics are passed through a decanter to recover the organic phase. The decanter shown in Figure 4-1 is recovering an organic phase lighter than water. The organic phase overflows the decanter and is normally pumped to a storage tank and then recycled to the process. When not recycled, the recovered organics are usually burned as fuel in an incinerator. The water phase in the decanter is recycled to the feed tank for feed to the stripper. In well designed systems, any noncondensable organic compounds are routed through the wastewater feed tank to a control device such as an incinerator or carbon adsorber. This strategy reduces the potential for air emissions from the steam stripper system. For systems where a primary condenser is not sufficient for condensing a large portion of the organics, secondary control may be required. A secondary condenser with brine or a refrigerant as the cooling agent can be installed to condense most of the organics. The condensates from the secondary condenser should be sent to a decanter, and the noncondensables should be vented to a control device.

4.2.1.3 Steam Stripper Removal Performance. The organic removal performance of five steam stripper systems was measured during field tests sponsored by the EPA. Information gathered during these testing efforts is summarized in Appendix C. During each test, samples of the feed and bottoms streams were obtained to determine the organic removal performance of the stripper. In addition, data were gathered to evaluate the effect of design and operating parameters on the performance of each system. These data, along with performance data provided by a facility using a recently installed steam stripper, are presented in Table 4-2. The organic removals presented in Table 4-2 range from 76 percent for Site 7 to greater than 99.8 percent for Site A. In general, the organic compound removal efficiencies are higher for the steam strippers treating wastewater containing chlorinated organic compounds. These chlorinated organic compounds are more easily steam stripped than organic compounds such as phenol that are more soluble and less volatile in water. The steam stripper at Site G, however, achieved a relatively high organic removal efficiency of 92 percent for moderately volatile nitrobenzene and nitrotoluene constituents. This efficiency is due in part to the column height for this stripper (19.2 m).

In some cases a facility may already be treating some of their existing wastewater streams with a steam stripper for product recovery or to meet other EPA standards such as the vinyl chloride standard. If a facility plans to use the existing stripper to control wastewaters from a major modification, the steam stripper system should achieve equal or better organic compound removal than the recommended steam stripper system. Similarly, if a facility installs

TABLE 4-2. STEAM STRIPPER ORGANIC COMPOUND REMOVAL PERFORMANCE

		Column Design			Column Operation		Column Performance				
							Feed		Bottoms		
Site I.D.	Organic Compounds	Height (m)	Diameter (m)	Feed Rate (kg/hr)	Steam: Feed Ratio (kg/kg)	Overheads: Feed Ratio (kg/kg)	Conc. (mg/l)	Loading (kg/hr)	Conc. (mg/l)	Loading (kg/hr)	Organic Compound Removal (%)
A	CHC, Benzene Chlorobenzene	NA	NA	48,960	0.03	0.03	5,900	290	9.8	0.5	>99.8
B	CHC	3.0[a]	0.2	1,260	0.1	0.01	3,900	4.7	5.2	0.005	99.8
G	Nitrobenzene Nitrotoluene	19.2	0.46	29,900	0.07	0.01	634	19	47.8	1.5	92
7	Benzene Nitrobenzene Phenol Nitrophenols	9.8[a]	0.61	5,452	0.4	NA	1,192	6.5	250	1.5	76
10	CHC Chlorinated ethers Phenol Chlorinated phenols	6.1[a]	1.07	12,693	NA	NA	453	5.7	3.4	0.04	99
Z	Benzene, Toluene Chlorobenzene Ethylbenzene Methylene chloride	12.2	1.22	68,100	0.020	0.0067	2,073	118	0.04	0.002	>99.8

[a]Height of packed section only. Total height is not available.

NA = Not available.

CHC = Chlorinated hydrocarbons.

a steam stripper other than the steam stripper design in Table 4-1 to treat wastewater streams from a new source, the steam stripper system should achieve equal or better removal than the recommended design. This emission reduction includes emissions which occur between the point of generation and the steam stripper, from the steam stripper vents, and those that occur downstream from the stripper. If a facility plans to use a steam stripper that does not meet the recommended design, additional removal efficiency might be achieved by adding packing or additional trays to the column if space in the column allows, or by increasing the steam flow rate. Another alternative would be to add a secondary treatment system such as a steam stripper system in series or liquid-phase carbon adsorption on the stripper effluent.

The removal efficiencies used in this document were predicted for the example waste stream composition using ASPEN. These data are presented in Table 4-1. As shown, the compounds with high Henry's Law constants were removed at efficiencies exceeding 99 percent. Figure 4-2 presents a curve developed to predict removal efficiency of any VOC based on the Henry's Law constant for the compound. A regression analysis was performed to relate removal efficiency as a function of Henry's Law constant for the five compounds in Table 4-1. The result was fraction removed = 1.357 + 0.08677 ln (Henry's Law constant, atm m^3/mol). (To get removal efficiency, multiply the fraction removed by 100.)

4.2.2 Other Organic Compound Removal Technologies

This section presents alternatives to steam stripping for reducing VOC emissions from industrial wastewaters. Although steam stripping is the most universally applicable technology for VOC emission reduction from industrial wastewaters, there are applications where another technology may be more appropriate. The purpose of this section is to present some of these alternatives along with a discussion of the technology.

In addition to steam stripping, technologies available for removing organic compounds from wastewater include air stripping, carbon and ion exchange adsorption, chemical oxidation, membrane separation, and liquid-liquid extraction.[34] These technologies rely on a variety of mechanisms to remove organic compounds from wastewater. However, with the exception of air stripping, the removal efficiencies of these technologies are dependent on

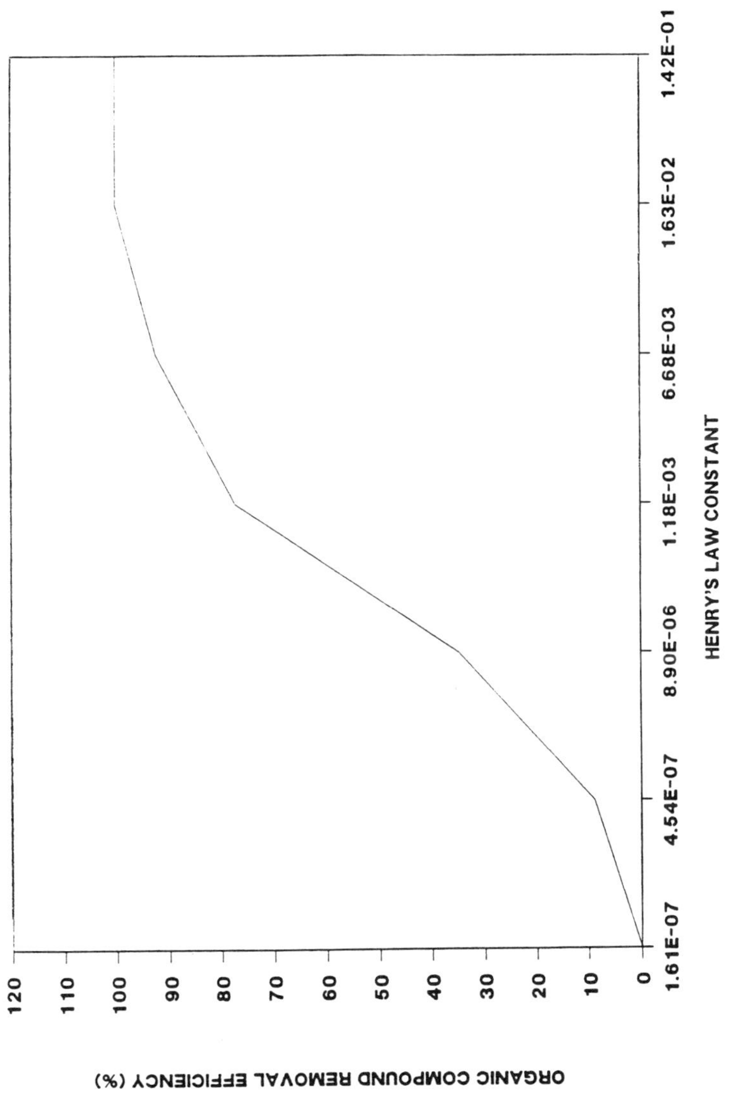

Figure 4-2. Predicted steam stripper organic compound removal efficiencies based on Henry's Law constant for the compound.

physical properties other than compound volatility in water. As discussed in Chapter 3, the organic compound property (volatility in water) has a major impact on air emissions during wastewater collection and treatment. Therefore, these technologies may not be as generally effective at reducing air emissions as steam strippers. However, these technologies are used in different applications by facilities in the targeted industries and may be effective at removing certain organic compounds. For this reason, a brief description of each technology is provided below.

The underlying principle for air stripping is vapor-liquid equilibrium.[35] By forcing large volumes of air through the contaminated water, the air-water interface is increased, resulting in an increase in the transfer rate of the organic compounds into the vapor phase. Although the technology is applicable to compounds with a wide range of volatilities, controlling the high volume of air needed to achieve high removal efficiencies on streams with large quantities organic compounds becomes prohibitive from a design and cost perspective. In addition, as the air rate through the column is increased, the cost of the VOC capture device also increases. In most cases, this capture device is a carbon adsorber. However, in some cases the air stream can be vented to a combustion device. In practice, air stripping is more applicable for streams containing dilute organic compound concentrations such as contaminated ground water.

Chemical oxidation involves a chemical reaction between the organic compounds and an oxidant such as ozone, hydrogen peroxide, permanganate or chlorine dioxide. The applicability of this technology depends on the reactivity of the individual organic compounds. For example, phenols and aldehydes are more reactive than alcohols and alkyl-substituted aromatics; halogenated hydrocarbons and saturated aliphatic compounds are the least reactive.[36]

Adsorption processes take advantage of compound affinities for a solid sorbent medium. Often activated carbon or polymeric resins are used as the medium. Nonpolar compounds can be adsorbed onto the surface of activated carbon. By contrast, removal by polymeric resins involves both adsorption and ion exchange mechanisms and is therefore, more effective for polar compounds. With carbon adsorption, the capacity of the carbon to adsorb the organic compounds at a given influent concentration varies widely for different compounds. In addition, the ease of desorption (removal) of the organic compounds and possible wastewater contaminants from the carbon is highly variable. For these reasons, the feasibility of using carbon adsorption must be evaluated on a case-by-case basis. A more detailed evaluation of the applicability of carbon adsorption to organic compound removal from industrial wastewaters is documented in a memorandum "Evaluation of Carbon Adsorption as a Control Technology for Reducing Volatile Organic Compounds (VOC) from Industrial Wastewaters" dated November 30, 1988.[37] Two types of membrane separation processes are: ultrafiltration and reverse osmosis. Ultrafiltration is primarily a physical sieving process driven by a pressure gradient across the membrane. This process separates macromolecular organic compounds with molecular weights of greater than 2,000, depending on the membrane pore size. Reverse osmosis is the process by which a solvent is forced across a semi-permeable membrane due to an osmotic pressure gradient. Selectivity is, therefore, based on osmotic diffusion properties of the compound and the sizes of the compound and membrane pores.[38]

Liquid-liquid extraction, sometimes referred to as solvent extraction, uses differences in solubility of compounds in various solvents as a separation technique. By contacting a solution containing the desired compound with a solvent in which the compound has a greater solubility, the compound may be removed from the solution. This technology is often used for product and process solvent recovery for two reasons. First, the solvent can usually be regenerated, and second, the compound of interest can often be recovered by distillation.

4.3 VOC EMISSION CONTROL FROM COLLECTION AND TREATMENT SYSTEM COMPONENTS

Within an industrial wastewater collection and treatment system, the individual components represent potential emission sources by providing contact between wastewater and ambient air. Through the use of physical covers and water seals, the contact between the wastewater and ambient air can be minimized, thus suppressing VOC emissions. Suppression controls can be broken down into four categories: collection system controls, roofs, floating membranes, and air-supported structures. Suppression controls are discussed in more detail in the following sections. However, suppressing these VOC emissions merely keeps the organic compounds in the wastewater until they reach the next potential VOC emission source. Therefore, these controls are not effective unless the VOC emissions are suppressed until the wastewater reaches a control device or a controlled treatment tank where the organic compounds are either removed or destroyed.

4.3.1 Collection System Controls

As discussed in Chapter 3, collection systems are comprised of components such as drains, junction boxes, sumps, trenches, and lift stations that provide contact between wastewater and ambient air. These collection system components provide escape routes for organic compounds contained in wastewater. Controls can be applied to most of these components to reduce the potential of VOC emissions during wastewater collection. These controls involve the use of physical covers and water seals to minimize the contact between ambient air and the wastewater flowing through the component. However, these types of controls serve merely to suppress VOC emissions; the organic compounds remain in the wastewater flowing downstream to the next potential emission source. For this reason, controls for these collection system components are only effective if the wastewater flows downstream to an organic compound removal or destruction device.

For this control strategy, the downstream removal devices are often oil/water separators and biological treatment units. If sufficient removal of organic compounds cannot be obtained in these components then an alternative removal technology should be investigated.

Figure 4-3 presents two commonly used methods for controlling emissions from drains: p-leg and seal pot configurations. In the p-leg configuration, water settles in the "P"-bend between the sewer line and the top of the drain riser. The seal pot configuration provides an external liquid seal. A cap covers the drain opening, and the bottom edge of the cap extends below the level of the drain entrance. Wastewater flows into the drain area outside the cap and then flows under the edge of the cap and into the drain. In both of these designs, the liquid seal prevents convective transfer of vapors from the sewer line below the drain. Vapors in the sewer must diffuse through the water seal to be emitted. Emission reductions for drains are addressed in the Background Information Document (BID) for VOC Emissions from Petroleum Refinery Wastewater Systems.[39] An emissions reduction study was performed at a refinery with drains equipped with seal pots having caps that could be manually removed. Screening results were evaluated for 76 drains both before

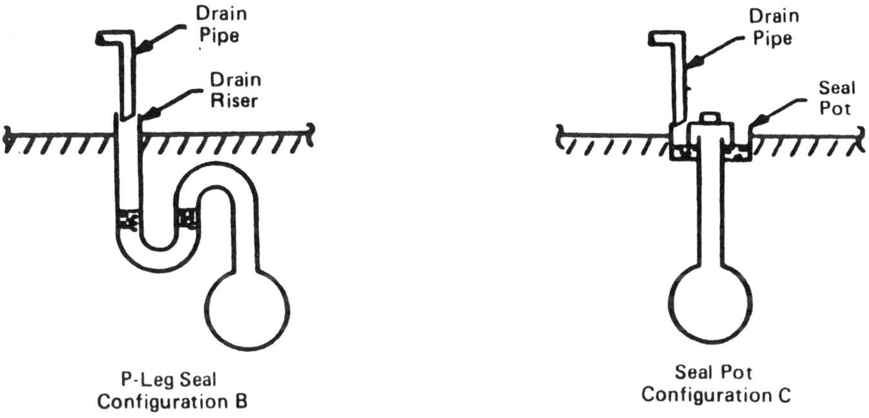

Figure 4-3. P-leg and seal pot configurations for drains.

and after the cap was removed. A further analysis grouped these drains into two categories to evaluate whether the uncontrolled leak rate had any effect on emission reduction. Those with uncontrolled screening values less than 100 ppm of organics were placed in one group while those with values greater than 100 ppm were placed in a second group. Of the 76 uncontrolled drains that were screened, 18 had leak rates greater than 100 ppm. Results of emission reduction vary from approximately zero to 99 percent. Disregarding outlying data points gives an average VOC emission reduction of approximately 95 percent for seal pots. (Data points were not considered for emission reductions less than zero or if screening values with the cap on were excessively high.) In cases where highly contaminated streams are continuously discharged to the drain or the organic compounds are immiscible and float on top of the water, the seal is ineffective. If this is the case, a completely enclosed drain system may be required. With these systems, the process drain lines are piped directly into the drains, which are connected to the sewer. A fuel gas purge then sweeps the VOC to a control device.

Use of p-leg seals and seal pots can reduce VOC emissions from drains if the system is well maintained; however, monitoring of the performance of the control will be difficult. Best control of emissions is achieved with hard piping any source of wastewater containing organic compounds to a control device.

Other collection system components that typically require control are junction boxes, sumps, and lift stations. Since the design of these three components are similar, the same technique is effective for controlling VOC emissions from all three. For these components, a gas tight cover is typically used. Figure 4-4 presents a schematic of a controlled junction box. As shown, a gas-tight cover is fitted on the top of the unit, with a vent line to the atmosphere. Although some VOC emissions do occur from this vent line, the losses are significantly less than from an uncontrolled unit. For junction boxes, sumps, and lift stations, a 95 percent control of VOC emissions is assumed with the application of a gas-tight cover.

Trenches are sometimes used to collect and transport wastewater from process areas. Since trenches are typically used for collection and transport of rainwater and wash-down water, as well as process wastewater, they must have openings and cannot be completely covered. Therefore, a reasonable technique for controlling emissions from trenches is to install a gas-tight cover with periodic openings along the length of the trench. These openings

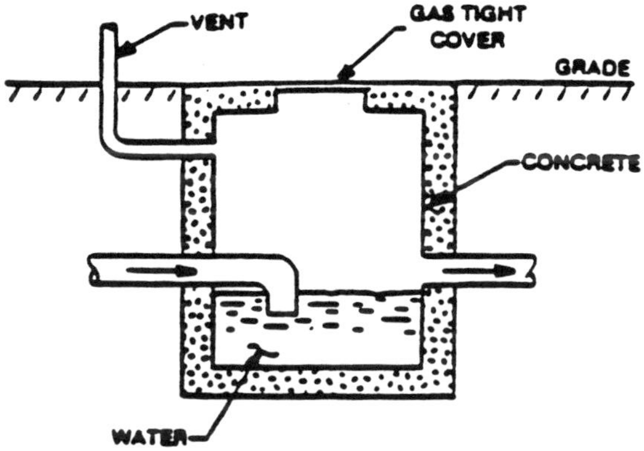

Figure 4-4. Gas tight cover for collection system components.

can then be controlled with a seal pot. As with junction boxes, sumps, and lift stations, a gas-tight cover applied to trenches is assumed to reduce VOC emissions by 95 percent.

4.3.2 Roofs

4.3.2.1 Fixed-Roof Tanks.[40] A fixed-roof tank is a vertical cylindrical steel wall tank with a cone-shaped or dome-shaped roof that is permanently attached to the tank shell (see Figure 4-5). Vents are installed on the roof to prevent the tank internal pressure from exceeding the tank design pressure limits and, thereby, causing physical damage or permanent deformation to the tank structure. The vents can either open directly to the atmosphere, be equipped with valves that open at specified pressure or vacuum settings, or be connected to an add-on control device (e.g., carbon adsorption system, vapor incinerator).

Storage or treatment of wastewater in fixed-roof tanks instead of open-top tanks reduces VOC emissions. By covering the tank, the wastewater surface is sheltered from the wind. This decreases the mass transfer rate of organic compounds in the wastewater to the atmosphere. The extent to which VOC emissions are reduced depends on many factors including wastewater composition and organic concentrations, windspeed, and the ratio of the tank diameter to the depth of the wastewater contained in the tank.

An existing open-top tank can be converted to a fixed-roof tank by retrofitting the tank with a dome roof. Aluminum, geodesic dome roofs are available from several manufacturers. These domes have been used successfully to cover petroleum and chemical storage tanks. The domes are clear-span, self-supported structures (i.e., require no internal columns be placed in the tank) that can be installed on open-top tanks ranging in diameter from 5 to over 100 m.

Although fixed-roof tanks provide large reductions in VOC emissions from open-top tanks, fixed-roof tanks still can emit significant quantities of VOC. The major sources of VOC emissions from fixed-roof tanks are breathing losses and working losses. Breathing losses occur from the expulsion of vapor through the roof vents because of the expansion or contraction of the tank vapor space resulting from daily changes in ambient temperature or barometric pressure. These VOC emissions occur in the absence of any liquid level change in the tank. Working losses occur from the displacement of vapors resulting

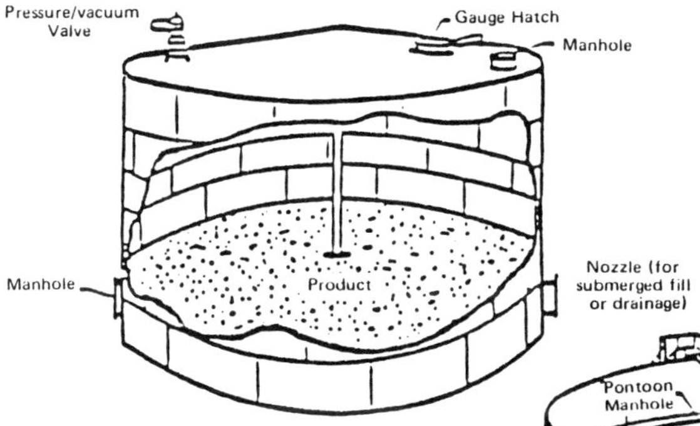

Typical fixed roof tank.

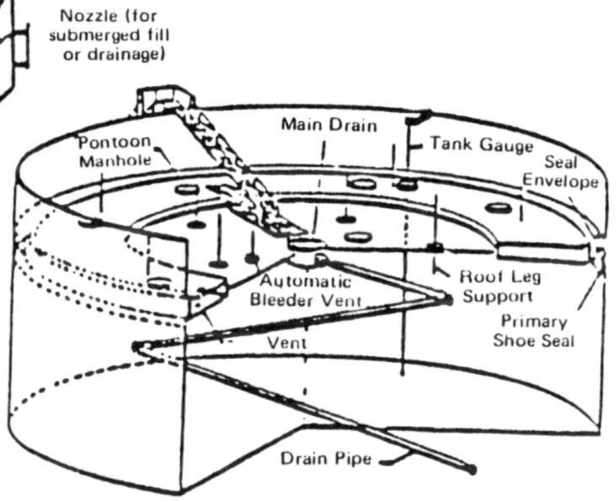

External floating roof tank.

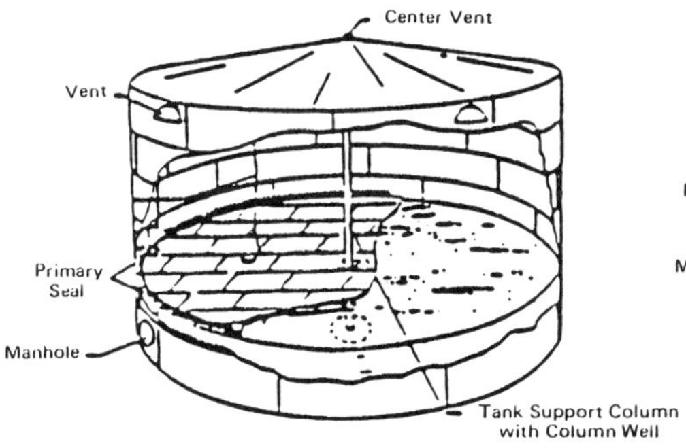

Contact Deck Type

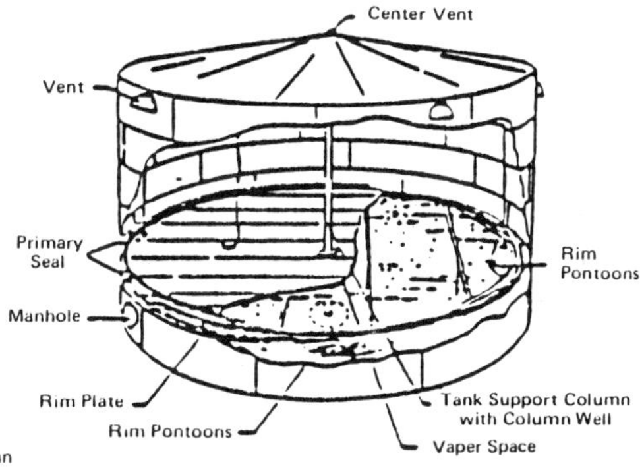

Noncontact Deck Type

Internal floating roof tanks

Figure 4-5. Storage tank covers.

from filling and emptying of the tank.

Breathing and working losses from fixed-roof tanks can be reduced by installing an internal floating roof, connecting the tank roof vents to an add-on control device, or installing pressure-vacuum relief valves on the tank roof vents. For add-on control applications, vapors are contained in the tank until the internal tank pressure attains a preselected level. Upon reaching this level, a pressure switch activates a blower to collect the vapors from the tank and transfer the vapors through piping to the add-on control device. As a safety precaution, flame arrestors normally are installed between the tank and control device. Other safety devices may be used such as a saturator unit to increase the vapor concentration above the upper explosive limit. Add-on control devices for organic vapors are discussed in Section 4.4.

4.3.2.2 Floating Roof Tanks.[41] Floating roofs are used extensively in the petroleum refining, gasoline marketing, and chemical manufacturing industries to control VOC emissions from tanks storing organic liquids. A floating roof is basically a disk-shaped structure (termed a "deck") with a diameter slightly less than the inside tank diameter that floats freely on the surface of the wastewater in the tank. A seal is attached around the outer rim of the deck to cover the open annular space between the deck and inside tank wall. The seal mechanism is designed to slide against the tank wall as the wastewater level in the tank is raised or lowered. There are two general types of tank floating roofs: external floating roofs and internal floating roofs.

Floating roofs are appropriate for wastewater storage tanks and certain treatment tanks where the presence of the floating cover would not interfere with the treatment process. Treatment tanks equipped with surface mixing or aeration equipment cannot use floating roofs. Also, because floating roofs are in direct contact with the wastewater, the materials selected to fabricate the deck and seals must be compatible with the wastewater composition.

An external floating roof consists of a single- or double-layer steel deck that moves within the walls of an open-top tank (see Figure 4-5). Pontoon sections often are added to the deck to improve floatation stability. Because the top surface of the deck is exposed to the outdoors, the external floating roof design must include additional components for rainwater drainage and snow removal to prevent the deck from sinking, and for cleaning the inside walls of the tank above the deck to protect the sliding seal mechanism from dirt. A variety of seal types (e.g., metallic shoe seal, liquid-filled seal, or resilient foam-filled seal) and seal configurations (e.g., mounted above liquid surface, mounted on liquid surface) can be used for external floating roofs. Small openings are required on the deck for various fittings such as vents, inspection hatches, gage wells, and sampling ports.

An internal floating roof consists of a steel, stainless steel, aluminum, or fiberglass-reinforced plastic deck that is installed inside a fixed-roof tank (see Figure 4-5). Many internal floating roof designs can be retrofitted into existing fixed-roof tanks. Because the fixed roof shelters the deck from weather, internal floating roofs do not need additional components for rainwater drainage or for seal protection. An internal floating roof is equipped with the same types of deck fittings used on an external floating roof, but normally uses a simpler deck seal mechanism (e.g., a single resilient foam-filled seal or wiper seal). Vertical guide rods are installed inside the tank to maintain deck alignment. The internal tank space above the deck must be vented to prevent the accumulation of a flammable vapor mixture.

Floating roof tanks significantly reduce but do not eliminate VOC emissions. Organic vapor losses termed "standing losses" occur at the deck seals and fitting openings. The imperfect fit of the deck seals allows gaps that expose a small amount of the liquid surface to the atmosphere. Small quantities of vapors that collect in the small openings under the deck can leak from the deck fitting openings. Standing losses can be reduced by installing secondary deck seals, selecting appropriate pressure-relief valve settings, and using tight-gasketed and bolted covers on all other fittings. Additional organic vapor losses termed "withdrawal losses" occur from evaporation of the liquid that wets the inside tank wall as the roof descends during emptying operations.

No emission source test studies of full-sized tanks equipped with floating roofs have been conducted because of the complexity of erecting an enclosure around a tank. However, emission test studies of full-sized floating roof components sponsored by the American Petroleum Institute (API) were conducted using a pilot-scale tank. The results of these studies in combination with other data have been used by API and EPA to develop empirical models that estimate external and internal floating roof tank standing and withdrawal losses.

For the development of volatile organic liquid storage New Source Performance Standards (NSPS), EPA analyzed the emission reduction effectiveness of using floating roof tanks compared to fixed-roof tanks using the empirical models. The percentage of reduction in emissions varies with the tank characteristics (e.g., tank size, vapor pressure of the material stored in the tank). A model tank was selected for the NSPS analysis that has a volume of 606 m^3, contains a volatile organic liquid having a vapor pressure of 6.9 kPa, and operates with 50 turnovers per year. The analysis concluded that, depending on the type of deck and seal system selected, installing an internal floating roof tank in a fixed-roof tank will reduce VOC emissions by 93 to 97 percent. The analysis also concluded that a similar level of emission reduction can be achieved using an external floating roof tank.

Tanks used for pH adjustment, equalization, decantation, and settling could use any of the tank types mentioned above if they are nonagitated. However, similar tanks and clarifiers which are agitated would be limited to fixed roofs. Any of the three tank types could be applied to oil/water separators. Oil/water separators are discussed in more detail below.

RCRA Subtitle D surface impoundments as described in Chapter 2 could apply floating membrane covers if they are nonagitated. Such impoundments might be used for evaporation, polishing, storage, equalization, etc. If these impoundments are agitated, they require air-supported structures erected over them. Surface impoundments and tanks used for biodegradation could apply the floating membrane covers if nonagitated or the air-supported structures if agitated.

In all cases the vapor space within the fixed roof or air-supported structure should be vented to an add-on control device. These devices are discussed in more detail in Section 4.4.

4.3.2.3 Oil/Water Separators. The most effective option for controlling VOC emissions from oil/water separators is to install either a fixed or floating roof. These roofs control VOC emissions by reducing the oil surface exposed to the atmosphere, reducing the effects of wind velocity, and reducing the effects of solar radiation by insulating the oil layer.

Fixed roofs can be installed on most oil/water separators. This can be done without interfering with the operation of the system by mounting on the sides of the separator or by supporting with horizontal steel beams set into the sides of the unit. Gas-tight access doors are usually installed in the roof for maintenance and inspection. Since the vapor space below fixed roofs may constitute an explosion or fire hazard, the vapor space is often blanketed with nitrogen and/or purged to a recovery or destruction device.

To eliminate the need for a nitrogen blanket or purge, floating roofs are sometimes installed on oil/water separators. Floating roofs may be constructed of plastic or glass foam blocks, aluminum pontoons, or fiberglass. To prevent the roof from interfering with operation of the flight scraper, the water level can be raised in the separator so that the top of the oil surface is above the flight scraper blades.

For both types of roofs, the effectiveness of their emission control is primarily dependent on the effectiveness of the seals between the roofs and walls of the separator. If these seals are not well-maintained to prevent leakage, their VOC emission control capabilities are reduced significantly. Emission reductions from covering separators are also limited by the method of controlling vent emissions. If the vent is directed to a control device rather than the atmosphere, greater emissions reduction will be achieved.

Although very little data are available regarding the VOC emission reduction achieved by oil/water separators, theoretical analyses have indicated that a floating roof can reduce emissions from the oil/water separator by at least 85 percent when equipped with a primary liquid seal and a secondary seal. Other sources report varying levels of emissions reduction but give no supporting documentation. The American Petroleum Institute (API) reports 90 percent to 98 percent reductions of emissions from the separator and an emission reduction of 96 percent is reported in the Compilation of Air Pollutant Emission Factors, AP-42. The State of California estimated 90 percent less emissions from the unit when equipped with a floating roof. Although it is dependent on numerous factors, including effectiveness of the control device to which the roof vent is directed, an efficiency of 90 percent is probably most reasonable for a fixed roof. Without any better information, 90 percent reduction of oil/water separator emissions should be assumed in estimating emissions from a fixed roof, vented to a control device, and 85 percent reduction for a floating roof.

One final concern in evaluating emissions from oil/water separators is the handling of the recovered oils. Since the oils may contain high concentrations of organic compounds, care must be taken to minimize VOC emissions. This can be accomplished by handling the oils and organics in closed systems equipped with emission controls as well.

Tables 4-3 and 4-4 present oil/water separator analyses performed on Example Waste Stream Schematic II. In Table 4-3, the oil/water separator is

TABLE 4-3. POTENTIAL VOC EMISSION REDUCTIONS THROUGH ENCLOSING OF COLLECTION SYSTEM AND COVERING OIL/WATER SEPARATOR WITH A FIXED ROOF VENTED TO A CONTROL DEVICE

	Butadiene	Toluene	Naphthalene	1-Butanol	Phenol	Total
Fraction Emitted Through the Oil/Water Separator[a]	0.11	0.049	0.0099	0.000090	0.0000050	0.034
Fraction Removed in Oil Layer[b]						
100 ppm oil	0.0065	0.022	0.32	0.00015	0.00014	0.070
1,000 ppm oil	0.061	0.18	0.69	0.0015	0.0014	0.19
10,000 ppm oil	0.34	0.57	0.78	0.015	0.014	0.35
Fraction Removed by Vent Control Device[c]						
100 ppm oil	0.88	0.41	0.055	0.00062	0.000034	0.27
1,000 ppm oil	0.83	0.35	0.025	0.00062	0.000034	0.24
10,000 ppm oil	0.54	0.18	0.018	0.00061	0.000033	0.15
Fraction Passthrough[d]						
100 ppm oil	0.00	0.52	0.61	0.999	0.9998	0.63
1,000 ppm oil	0.00	0.43	0.27	0.998	0.999	0.54
10,000 ppm oil	0.00	0.20	0.19	0.98	0.99	0.47

[a] Assumes 90 percent reduction of emission from oil/water separator, and 99 percent reduction of emission from collection system components.
[b] Assumes 80 percent of the organic compounds that partition into the oil are removed.
[c] Assumes 90 percent of organic compounds that would be emitted in an uncontrolled oil/water separator are removed.
[d] Remaining organic compounds pass on to next potential emission point.

TABLE 4-4. POTENTIAL VOC EMISSION REDUCTIONS THROUGH ENCLOSING OF COLLECTION SYSTEM AND COVERING OIL/WATER SEPARATOR WITH A FLOATING ROOF

	Butadiene	Toluene	Naphthalene	1-Butanol	Phenol	Total
Fraction Emitted Through the Oil/Water Separator[a]	0.16	0.072	0.014	0.00012	0.0000068	0.050
Fraction Removed in Oil Layer[b]						
100 ppm oil	0.0062	0.022	0.32	0.00015	0.00014	0.070
1,000 ppm oil	0.057	0.17	0.69	0.0015	0.0014	0.18
10,000 ppm oil	0.33	0.56	0.78	0.015	0.014	0.34
Fraction Pass through[c]						
100 ppm oil	0.83	0.91	0.66	0.9997	0.9998	0.88
1,000 ppm oil	0.78	0.76	0.30	0.999	0.999	0.77
10,000 ppm oil	0.51	0.37	0.21	0.98	0.99	0.61

[a]Assumes 85 percent reduction of emission from oil/water separator, and 99 percent reduction of emission from collection system components.

[b]Assumes 80 percent of the organics that partition into the oil are removed.

[c]Remaining organic compounds pass on to next potential emission point.

assumed to be controlled with a fixed roof vented to a control device. Three cases are presented for each of the five representative compounds: 100 ppm, 1,000 ppm, and 10,000 ppm of oil in the wastewater. When an oil/water separator is covered with a fixed roof and vented to a control device, removal of organic compounds is achieved in two ways: (1) Organic compounds that partition into the oil layer are removed with this oil; and (2) Organic compounds that are volatilized and swept through a nitrogen purge to a vent control device are recovered or destroyed. For the purpose of this analysis, 99 percent emission reduction was assumed for each collection system component, and 90 percent of the potential emissions in the oil/water separator are assumed to be recovered or destroyed in a control device, with the remaining 10 percent of these potential emissions being emitted.

As shown in Table 4-3, the compounds' vapor pressure and solubility, and the amount of the organics partitioning into the oil layer (a function of the octanol-water partition coefficient for the compound, and the total oil fraction in the water) have significant impacts on the potential removal. Compounds such as butadiene may be removed reasonably efficiently (54 to 88 percent) by the control device on the oil/water separator vent if they are controlled efficiently up to the separator. As shown, 11 percent of the butadiene was emitted prior to removal and a total of 89 percent was removed by the two mechanisms. Other compounds, such as naphthalene partition into the oil layer so completely that they can be removed efficiently only if the oil fraction is sufficiently large. Removal of this compound in the oil ranged from 32 to 78 percent, depending on the oil concentration. Still other compounds pass through to the next treatment unit. If these compounds' emissions are not controlled effectively downstream, very little overall emission reduction is achieved.

Table 4-4 presents the same scenario as Table 4-3 except the fixed roof is replaced by a floating roof. Floating roofs are assumed to reduce emissions by 85 percent but are not vented to a control device. The only removal mechanism in this case is the removal of the organics that partition into the oil layer and are removed with the oil. As shown, for most compounds, this control primarily suppresses the VOC emissions and unless adequate controls are present downstream, the VOC will be emitted. Of the five compounds presented, greater than 56 percent removal is only attained for naphthalene. Naphthalene removal ranges from 32 percent to 78 percent depending on the oil fraction.

4.3.3 Floating Membrane Covers[42]

A floating membrane cover consists of large sheets of a synthetic flexible membrane material that floats on the surface of the wastewater. Individual sheets can be seamed or welded together to form covers applicable to any size area. Floating membrane covers have been used successfully for many years to cover the surface of potable water impoundments or reservoirs. In a "leak tight" application, floating membrane covers have been used to cover large anaerobic digester lagoons to collect the methane gas for energy recovery. Thus, floating membrane covers offer good potential as a suppression device for wastewater surface impoundments.

The material used to fabricate a floating membrane cover for wastewater unit applications needs to be resistant to chemical and biological degradation from compounds in the wastewater while also having good strength characteristics to resist tearing and puncturing. Synthetic materials used successfully for hazardous waste landfill bottom liners such as high-density polyethylene (HDPE) or polyvinyl chloride (PVC) are good candidate materials for floating membrane covers. The material currently preferred by most floating membrane cover vendors is HDPE.

Although HDPE is buoyant in water, foam floats are normally placed under the membrane to provide additional flotation. To prevent sinking of the cover because of accumulation of rainfall on top of the cover, the cover is designed with sufficient excess material to form troughs that collect rainfall. These troughs are connected to a pump system. Automatic controls periodically activate the pumps to drain the accumulated water off the cover. Pressure-relief valves are also likely to be installed on the cover to provide emergency venting of any gas buildup under the cover.

The simplest and least expensive method for anchoring a floating membrane cover is to dig a trench around the impoundment perimeter, insert the edge of the cover into the trench, and then backfill the trench with earth to secure the cover. An alternative but more expensive method is to construct a continuous concrete grade-level footing or short wall around the perimeter of the impoundment. The cover is then mechanically fastened to the top of the footing or wall using gaskets for tightness. This method has been successfully used for the anaerobic digester lagoon applications.

Application of floating membrane covers to wastewater surface impoundments will require special provisions for impoundment cleaning.

Floating membrane covers for surface impoundments are heavy, weighing over 2 kg/m. Consequently, once the cover is installed, removal of the entire cover is difficult. Therefore, a section of the cover will need to be removed and a temporary bracing system devised to provide access to the impoundment by the cleaning equipment.

The effectiveness of a floating membrane cover depends on the amount of wastewater surface that is covered and the permeability of the membrane material to the organic compounds contained in the wastewater. Using a membrane material with adequate thickness and following good installation practices will minimize tearing or puncturing the membrane material. Permeation of the cover is a three-step process that involves the absorption of the organics by the membrane material, diffusion of the organics through the membrane, and evaporation of the organics on the air side of the membrane. The overall cover permeability is a function of the organic composition and concentration of the wastewater managed in the surface impoundment as well as the cover material's composition and thickness.

Benchscale study of the floating membrane is ongoing. Laboratory tests are being performed with different compounds to determine diffusion rates through the membrane. The benchscale model is set up to be essentially leak proof. Therefore, diffusion through the membrane is the primary VOC emission point. Preliminary results show that for some compounds diffusion through the membrane can be fairly rapid. However, additional results are needed to provide a complete and accurate evaluation of the method. For the present, overall suppression efficiency of floating membrane covers applied to hazardous waste surface impoundments is estimated to be 85 percent.

4.3.4 Air-Supported Structures[43]

An air-supported structure is a plastic-reinforced fabric shell that is inflated and, therefore, requires no internal rigid supports. Figure 4-6 shows the major air-supported structure components. The structure shape and support is provided by maintaining a positive interior pressure (i.e., the interior pressure is greater than the external atmospheric pressure).

Large electric-motor driven fans are used to blow air continuously or intermittently through the structure and out a vent system. The interior pressure is maintained at a constant 10 to 15 kPa for structure inflation.

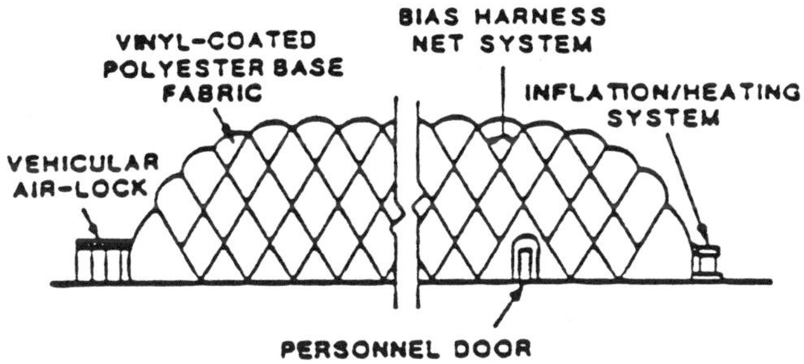

Source: Air Structures International, Inc.

Figure 4-6. Typical air-supported structure.

Adequate air changes are necessary to prevent the organic vapor concentrations inside the structure from exceeding the lower explosive limits. A standby blower system consisting of internal combustion engine driven fans normally is installed to keep the structure inflated and ventilated in the event of an electrical power outage. The vent system can discharge directly to the atmosphere or be connected to an add-on control device.

Large areas can be enclosed by erecting an air-supported structure. Structures are commercially available ranging in widths from 24 to 91 m wide and lengths from 24 to 137 m. For larger areas, a number of modules can be connected together. Air-supported structures have been used as enclosures for conveyors and coke ovens, open-top tanks, and material storage piles. A 4,000 m^3 aerated wastewater treatment lagoon at a specialty chemical manufacturing plant has been covered by an air-supported structure for more than four years. Thus, air-supported structures offer good potential as a suppression device for wastewater surface impoundments that cannot use floating membrane covers (e.g., surface treatment impoundments using surface-mounted aeration equipment).

The fabric used for the air-supported structure depends on the size of the structure, design requirements (e.g., wind and snow loadings), and type of chemicals to which the fabric's inner side will be exposed. Polyvinyl-chloride-coated polyester fabric would likely be the material of current choice for wastewater treatment applications because of the fabric's good resistance to deterioration from chemical, weather, or ultraviolet sunlight exposure. The service life of the fabric ranges from 2 to 12 years depending on the site-specific conditions.

Anchoring the air-supported structure likely will be accomplished by bolting the edges of the fabric to a continuous, grade-level concrete footing or beam installed around the perimeter of the surface impoundment. Entrance into an air-supported structure is through airlocked doors. These doors can be sized to allow earth-moving equipment to be used inside the structure for impoundment cleaning operations.

The use of air-supported structures to enclose wastewater impoundments can result in excessive condensation and high temperatures inside the structure. An air-supported structure's interior temperatures typically are 5 to 11°C above the ambient temperature. Consequently, during hot summer days, temperatures inside an air-supported structure can exceed 42°C. Depending on the severity of these conditions, workers entering the structure may need to follow additional safety procedures and be restricted as to the period of time they may remain inside the structure. Also, any equipment operating inside the structure may require more frequent repair or replacement because of accelerated rust and corrosion of the equipment components.

The effectiveness of an air-supported structure in suppressing VOC emissions primarily depends on the amount of leakage from the structure and whether the structure vent system is connected to an add-on control device. Air-supported structure leaks are usually confined to areas around airlocks, doors, and anchor points. Leak checks were performed at the air-supported structure operating at the specialty chemical manufacturing plant. A soap solution was sprayed around the structure base and fittings to locate leaks, and measurements were made using a portable hydrocarbon analyzer. Few leaks were found, and the sizes of the leaks ranged from 20 to 40 ppm. The operating experience at this facility indicates that proper installation and maintenance of the air-supported structure can limit leakage to very low levels.

Because of the low leakage levels attainable, almost all of the organic vapors contained by an air-supported structure will be ultimately discharged through the structure's vent system. Therefore, connecting the vent system to one of the add-on control devices discussed in Section 4.4 will result in an overall VOC emission control efficiency for wastewater treatment applications using an air-supported structure that is approximately equivalent to the efficiency of the control device. These add-on control devices are capable of achieving control efficiencies in excess of 95 percent.

Operation of an air-supported structure consumes large quantities of electricity to maintain the positive interior pressure. For example, the existing air-supported structure covering a 4,000 m^3 aerated wastewater treatment lagoon uses fans with a combined power rating of 26 kW for structure inflation and ventilation. Annual electricity consumption to operate continuously a standard 26 kW fan is approximately 250,000 kWh. Application of air-supported structures to wastewater emission source increases demand for electricity.

4.4 ADD-ON CONTROLS

Add-on controls are processes applied to capture organic vapors vented from wastewater emission sources. These controls serve to reduce VOC emissions by destroying organics in the gas stream or extracting organics from the gas stream before discharging the gas stream to the atmosphere. Add-on controls for VOC emissions are classified into four broad categories: combustion, adsorption, condensation, and absorption. General background information about these types of add-on controls is available in Reference 44. The type of add-on control best suited for a particular wastewater emission source depends on the size of the source and the characteristics of the wastewater in the source.

Combustion destroys the organics in the gas stream by oxidation of the compounds to primarily carbon dioxide and water. Because essentially all organics will burn, combustion add-on controls are applicable to all emission sources for which the organic vapors can be captured. Combustion add-on controls are thermal vapor incinerators, catalytic vapor incinerators, flares, boilers, and process heaters.

4.4.1 Carbon Adsorbers[45]

Adsorption as applied to air pollution control is the process by which organic molecules in a gas stream are retained on the surface of solid particles. The solid most frequently used is carbon that has been processed or "activated" to have a porous structure. This provides many surfaces upon which the organic molecules can attach, resulting in a high rate of organic removal from a gas stream as it passes through a bed of carbon.

Activated carbon has a finite adsorption capacity. When the carbon becomes saturated (i.e., all of the carbon surface is covered with organic material), there is no further VOC emission control because all of the organic vapors pass through the carbon bed. At this point (referred to as "breakthrough"), the organic compounds must be removed from the carbon before VOC emission control can resume. This process is called desorption or regeneration.

For most air pollution control applications, regeneration of the carbon in the adsorber is performed by passing steam through the carbon bed. The steam heats the carbon particles, which releases the organic molecules into the steam flow. The resulting steam and organic vapor mixture is condensed to recover the organics and separate the water for discharge to a wastewater treatment unit. An alternative method for regenerating the carbon is to reduce the pressure of the atmosphere surrounding the carbon particles.

Vacuum regeneration is used for special carbon adsorber applications when direct recycling of the recovered organics is desired such as vapor recovery at gasoline tank truck loading terminals. A detailed description of carbon adsorption and desorption mechanisms is available in Reference 45.

Two types of carbon adsorption systems most commonly used for VOC emission control are: fixed-bed carbon adsorbers and carbon canisters. A fluidized-bed carbon adsorption system has been developed but currently is not commercially available.

Fixed-bed carbon adsorbers are used for controlling continuous, organic gas streams with flow rates ranging from 30 to over 3,000 m^3/min. The organic concentration can be as low as several parts per billion by volume (ppbv) or as high as 25 percent of the lower explosive limit of the vapor stream constituents. The major components of a fixed-bed carbon adsorber system are one or more carbon bed units to adsorb the organics, a condenser to convert the desorbed organics and a steam mixture to a liquid, a decanter to separate the organic and aqueous phases, and blowers to cool and dry the carbon beds following desorption.

Fixed-bed carbon adsorbers may be operated in either intermittent or continuous modes. For intermittent operation, the adsorber removes organics only during a specific period of the day. Intermittent mode of operation allows a single carbon bed to be used because it can be regenerated during the off-line periods. For continuous operation, the unit is equipped with two or more carbon beds so that at least one bed is always available for adsorption while other beds are being regenerated.

Carbon canisters differ from fixed-bed carbon adsorbers. First, a carbon canister is a very simple add-on control device consisting of a 0.21 m^3 drum with inlet and outlet pipe fittings (see Figure 4-7). A typical canister unit is filled with 70 to 90 kg of activated carbon. Second, use of carbon canisters is limited to controlling low volume gas streams with flow rates less than 3 m^3/min. Third, the carbon cannot be regenerated directly in the canister. Once the activated carbon in the canister becomes saturated by the organic vapors, the carbon canister must be removed and replaced with a fresh carbon canister. The spent carbon canister is then recycled or discarded depending on site-specific factors.

The design of a carbon adsorption system depends on the inlet gas stream characteristics including organic composition and concentrations, flow rate, and temperature. Good carbon adsorber performance requires that (1) the

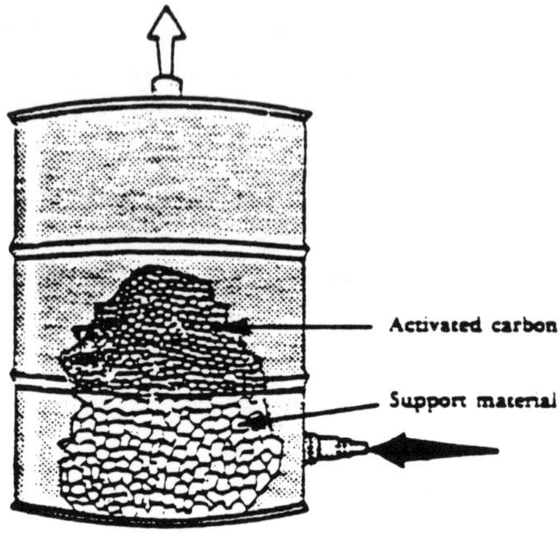

Figure 4-7. Carbon canister unit.

adsorber is charged with an adequate quantity of high-quality activated carbon; (2) the gas stream receives appropriate preconditioning (e.g., cooling, filtering) before entering the carbon bed; and (3) the carbon beds are regenerated before breakthrough occurs.

Emission source test data for 12 full-sized, fixed-bed carbon adsorbers operating in industrial applications has been compiled by EPA for a study of carbon adsorber performance.[46] The analysis of these data concluded that for well-designed and operated carbon adsorbers continuous organic removal efficiencies of at least 95 percent are achievable over long periods. Several units have been shown to continuously achieve organic removal efficiencies of 97 to 99 percent.

An equivalent level of performance for carbon canisters applied to a wastewater treatment unit is indicated by the results of an emission source test conducted on carbon canisters installed on the neutralizer tanks for a wastewater treatment system at a specialty chemicals plant.[47] This device was designed for odor control and not organic removal. However, 100 percent removal was measured for 1,2-dichlorobenzene, benzene, toluene, chlorobenzene, and chloroform. Overall organic removal efficiencies measured for various hydrocarbon categories ranged from 50 to 99 percent.

High moisture content in the gas stream can affect carbon adsorber performance for gas streams having organic concentrations less than 1,000 ppm.[48] At these conditions, water molecules compete with the organic compounds for the available adsorption sites on the carbon particles. Consequently, the carbon bed working capacity is decreased. Above an organic concentration of 1,000 ppm, high moisture does not significantly affect performance. Thus, to obtain good adsorber performance for gas streams with a high relative humidity (relative humidity greater than 50 percent) and low organic concentration (less than 1,000 ppm) requires preconditioning the gas stream upstream of the carbon bed. This can be accomplished using a dehumidification system, installing duct burners to heat the gas stream, or diluting the gas stream with ambient air. These gas stream conditions would most likely occur at locations where a carbon adsorber is used in conjunction with an air-supported structure enclosing an aerated surface impoundment containing dilute aqueous hazardous waste.

Carbon bed operating temperature can also affect carbon adsorber performance. Excessive bed temperatures can result due to the release of heat from exothermic chemical reactions that may occur in the carbon bed.[49]

Ketones and aldehydes are especially reactive compounds that exothermically polymerize in the carbon bed. If temperatures rise too high, spontaneous combustion will result in carbon bed fires. To avoid this problem, carbon adsorbers applied to gas streams containing these types of compounds must be carefully designed and operated to allow sufficient airflow through the bed to remove excess heat.

4.4.2 Thermal Vapor Incinerators[50]

Thermal vapor incineration is a controlled oxidation process that occurs in an enclosed chamber. Figure 4-8 shows a simplified diagram of a thermal vapor incinerator. One type of thermal vapor incinerator consists of a refractory-lined chamber containing one or more discrete burners that premix the organic vapor gas stream with the combustion air and any required supplemental fuel. A second type of incinerator uses a plate-type burner firing natural gas to produce a flame zone through which the organic vapor gas stream passes. Packaged thermal vapor incinerators are commercially available in sizes capable of handling gas stream flow rates ranging from approximately 8 to 1,400 m^3/min.[51]

Organic vapor destruction efficiency for a thermal vapor incinerator is a function of the organic vapor composition and concentration, combustion zone temperature, the period of time the organics remain in the combustion zone (referred to as "residence time"), and the degree of turbulent mixing in the combustion zone. Test results and combustion kinetics analyses indicated that thermal vapor incineration destroys at least 98 percent of non-halogenated organic compounds in the vapor stream at a temperature of 870°C and achieves a residence time of 0.75 seconds.[36] If the vapor stream contains halogenated compounds, a temperature of 1,100°C (2,000°F) and a residence time of one second is needed to achieve a 98 percent destruction efficiency.[52]

Incinerator performance is affected by the heating value and moisture content of the organic vapor stream, and the amount of excess combustion air. Combustion of organic vapor streams with a heating value less than 1.9 MJ/m^3 usually requires the addition of supplemental fuel (also referred to as auxiliary fuel) to maintain the desired combustion temperature.[53] Above this heating value, supplemental fuel may be used to maintain flame stability.

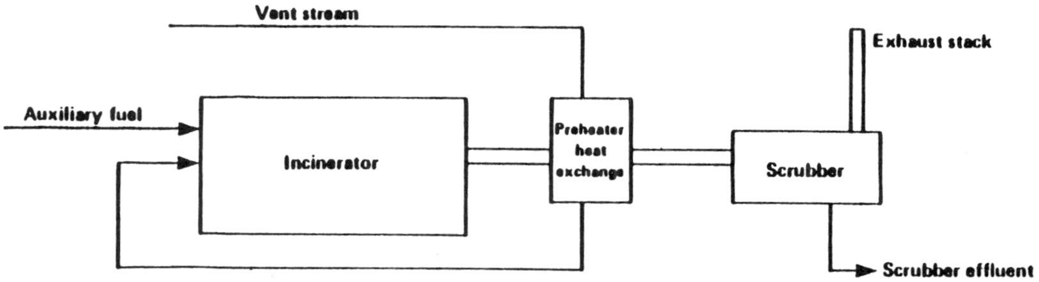

Figure 4-8. Schematic diagram of thermal incinerator system.

Although either natural gas or fuel oil can be used as supplemental fuel, natural gas is preferred. Supplemental fuel requirements can be decreased if the combustion air or organic vapor stream is preheated.

4.4.3 Catalytic Vapor Incinerators[54]

Catalytic vapor incineration is essentially a flameless combustion process. Passing the organic vapor stream through a catalyst bed promotes oxidation of the organics at temperatures in the range of 320 to 650°C. Temperatures below this range slow down or stop the oxidation reactions resulting in low destruction efficiencies. Temperatures above this range shorten catalyst life or may even cause catalyst failure. Oxidation of vapor streams with a high organic content can produce temperatures well above 650°C. Consequently, high organic concentration vapor streams may not be suitable for catalytic incineration. Figure 4-9 shows a simplified diagram of a catalytic vapor incinerator. The device consists of a chamber where the gas stream vented from the emission source is heated to the desired reaction temperature by mixing the organic vapors with hot combustion gas from natural gas-fired burners. The heated gas mixture then flows through the catalyst bed. The catalyst is composed of a porous inert substrate material that is plated with a metal alloy containing platinum, palladium, copper, chromium, or cobalt. A heat exchange is installed to preheat the vapor stream and, hence, reduce the amount of fuel that must be burned.

Organic vapor destruction efficiency for catalytic vapor incinerators is a function of organic vapor composition and concentration, catalyst operating temperature, oxygen concentration, catalyst characteristics, and the ratio of the volumetric flow of gas entering the catalyst bed to the volume of the catalyst bed (referred to as "space velocity"). Destruction efficiency is increased by decreasing the space velocity. However, a lower space velocity increases the size of the catalyst bed and, consequently, the incinerator capital cost. For a specific catalyst bed size, increasing the catalyst bed temperature allows a higher space velocity to be used without impairing destruction efficiency.

A series of studies has been sponsored by EPA to investigate the destruction efficiency of catalytic vapor incinerators used to control organic and hazardous air pollutants. The results of these studies concluded that destruction efficiencies of 97 to 98 percent are achievable.

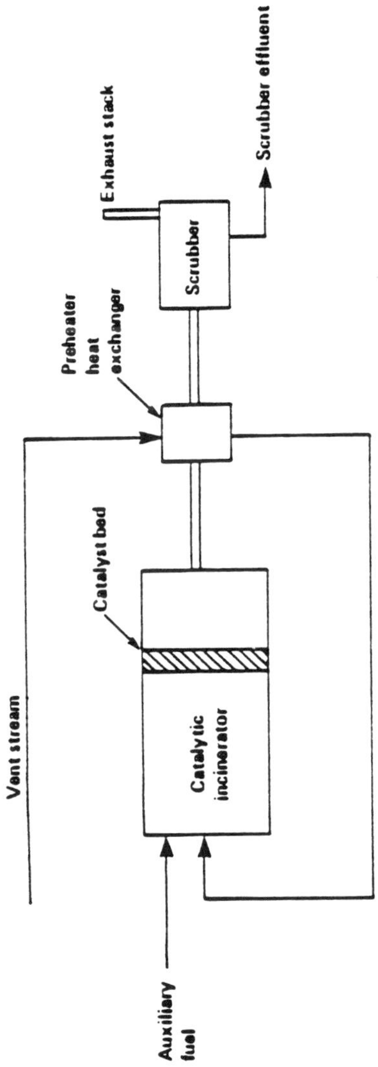

Figure 4-9. Schematic diagram of catalytic incinerator system.

The destruction efficiency is reduced by the accumulation of particulate matter, condensed organics, or polymerized hydrocarbons on the catalyst. These materials deactivate the catalyst by permanently blocking the active sites on the catalyst surface. If the catalyst is deactivated, the volatile organic compounds in the gas stream will pass through the catalyst bed unreacted or form new compounds such as aldehydes, ketones, and organic acids. Catalysts can also be deactivated by compounds containing sulfur, bismuth, phosphorous, arsenic, antimony, mercury, lead, zinc, tin, or halogens.

4.4.4 Flares[55]

Unlike vapor incinerators, a flare is an open combustion process. The ambient air surrounding the flare provides the oxygen needed for combustion. Consequently, a flare does not require blowers to provide combustion air. To achieve smokeless flare operation, turbulent mixing of the organic vapor stream with the ambient air at the flame zone boundary can be "assisted" by injecting steam or air at the flare tip or by releasing the gas stream through a high velocity nozzle (i.e., a nozzle with a high pressure drop). Flares are used extensively to burn purge and waste gases from many industrial processes such as petroleum refinery process units, blast furnaces and coke ovens.

Figure 4-10 shows a diagram of a typical steam-assisted flare configuration. The knockout drum is used to remove entrained liquids from the organic vapor stream. A water seal is used to prevent air intrusion into the flare stack. A pilot burner fired with natural gas is used to ignite the waste gases.

Flares without assist continuously burn the vapors from the emission source. A flare equipped with a steam, air, or pressure assist operates on an intermittent basis. Steam-assisted flares typically are used for burning large volumes of waste gases released from a process unit during upset or emergency condition. Air-assisted flares are less expensive to operate than steam-assisted flares. However, air-assisted flares are not suitable for large gas volumes because the airflow is difficult to control when the gas flow is intermittent. Pressure-assisted flares normally are used for applications requiring ground-level operation.

A series of flare destruction efficiency studies has been performed by EPA. Based on the results of these studies, EPA concluded that 98 percent combustion efficiency can be achieved by steam-assisted and air-assisted

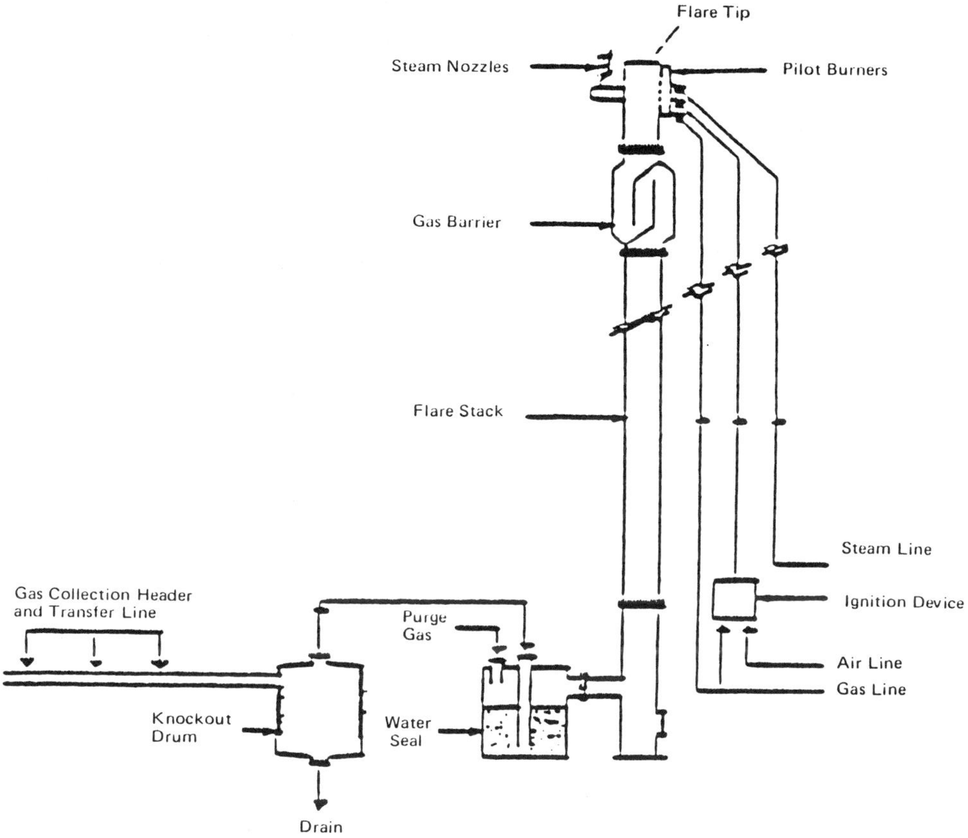

Figure 4-10. Steam-assisted elevated flare system.

flares burning gases with heat contents greater than 11 MJ/m^3. To achieve this efficiency level, EPA developed a set of flare design guidelines. The guidelines specify flare tip exit velocities for different flare types and waste gas stream heating values.

4.4.5 Boilers and Process Heaters[56]

A boiler or process heater can be used for organic vapor destruction. The organic vapor stream is either (1) premixed with a gaseous fuel and fired using the existing burner configuration, or (2) fired separately through a special burner or burners that are retrofitted to the combustion unit. Industrial boilers and process heaters currently are being used to burn vent gases from chemical manufacturing, petroleum refining, and pulp and paper manufacturing process units.

A series of EPA-sponsored studies of organic vapor destruction efficiencies for industrial boilers and process heaters was conducted by premixing waste materials with the fuel used to fire representative types of combustion devices. The destruction efficiency was determined based on the waste constituent concentrations measured in the fuel feed and stack gases using a gas chromatograph. The results of one study indicated that the destruction efficiency for an industrial boiler firing fuel oil spiked with polychlorinated biphenyls (PCB) was greater than 99.9 percent. A second study investigated the destruction efficiency of five process heaters firing a benzene vapor and natural gas mixture. The results of these tests showed 98 to 99 percent overall destruction efficiencies for C_1 to C_6 hydrocarbons.

Industrial boilers and process heaters are located at a plant site to provide steam or heat for a manufacturing process. Because plant operation requires these combustion units to be on-line, boilers and process heaters are suitable for controlling only organic vapor streams that do not impair the combustion device performance (e.g., reduce steam output) or reliability (e.g., cause premature boiler tube failure).

4.4.6 Condensers[57]

Condensation is the process by which a gas or vapor is converted to a liquid form by lowering the temperature or increasing the pressure. This process occurs when the partial pressure for a specific organic compound in the vapor stream equals its partial pressure as a pure substance at operating conditions. For air pollutant control applications, cooling the gas stream is the more cost-effective method of achieving organic condensation.

There are two major types of condensers: surface condensers and contact condensers. In a surface condenser the coolant does not contact the vapors or the condensate. In a contact condenser the coolant and vapor stream are physically mixed together inside the vessel and exit the condenser as a single stream.

A shell-and-tube-type heat exchanger is used for most surface condenser applications (see Figure 4-11). The gas stream flows into a cylindrical shell and condenses on the outer surface of tubes that are chilled by a coolant flowing inside the tubes. The coolant used depends on the saturation temperature or dewpoint of the particular organic compounds in the gas stream. The condensed organic liquids are pumped to a tank. Additional information about condenser equipment and operations is available in Reference 58.

The organic compound removal efficiency for a condenser is dependent upon the gas stream organic composition and concentrations as well as the condenser operating temperature. Condensation can be an effective control device for gas streams having high concentrations of organic compounds with high-boiling points. However, condensation is not effective for gas streams containing low organic concentrations or composed primarily of low-boiling point organics. At these conditions, organics cannot readily be condensed at normal condenser operating temperatures.

A field evaluation of a condenser used to recover organics from a steam stripping process used to treat wastewater at a plant manufacturing ethylene dichloride and vinyl chloride monomer was conducted. The measured condenser removal efficiencies for specific organic constituents ranged from a high value of 99.5 percent for 1,2-dichloroethane to a low value of six percent for vinyl chloride.[59]

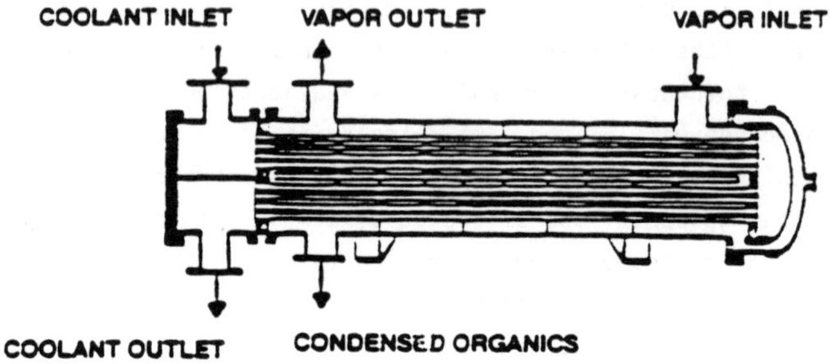

Figure 4-11. Schematic diagram of a shell-and-tube surface condenser.

4.5 REFERENCES

1. Letter and attachments from Radian Corporation to P. E. Lassiter, EPA/OAQPS, July 17, 1987, on the summary of data provided in confidential Section 114 questionnaire responses.

2. Perry, R. H. (ed.), Chemical Engineers' Handbook, 5th ed. New York, McGraw-Hill Book Co., 1973. pp. 13-50 - 13-55.

3. U. S. Environmental Protection Agency/ORD/HWERL. Preliminary Assessment of Hazardous Waste Pretreatment as an Air Pollution Control Technique. EPA-600/2-86/028 (NTIS PB46-17209/A6), March 1986.

4. Reference 1.

5. Trip Report. R. H. Howle and M. A. Vancil, Radian Corporation, to file. 7 p. Report of May 12, 1987, visit to Allied Fibers.

6. Trip Report. D. J. Herndon and S. K. Buchanan, Radian Corporation, to file. 10 p. Report of May 22, 1987, visit to Rhone-Poulenc Agricultural (RP Ag) Company.

7. Trip Report. D. J. Herndon and S. K. Buchanan, Radian Corporation, to file. 5 p. Report of September 2, 1987, visit to Fritzsche Dodge & Olcott.

8. Trip Report. D. J. Herndon, Radian Corporation, to file. 5 p. Report of May 6, 1987, visit to PPG Industries.

9. Trip Report. D. J. Herndon and S. K. Buchanan, Radian Corporation, to file. 7 p. Report of May 21, 1987, visit to Mobay Chemical Company.

10. Trip Report. D. J. Herndon, Radian Corporation, to file. 11 p. Report of May 4, 1987, visit to Borden Chemical Company.

11. Trip Report. D. J. Herndon, Radian Corporation, to file. 5 p. Report of May 6, 1987, visit to PPG Industries.

12. Trip Report. D. J. Herndon, Radian Corporation, to file. 4 p. Report May 5, 1987, visit to Union Carbide Corporation.

13. Trip Report. D. J. Herndon, Radian Corporation, to file. 13 p. Report May 8, 1987, visit to Dow Chemical Company.

14. U. S. Environmental Protection Agency. Hazardous Waste Treatment, Storage, and Disposal Facilities (TSDF) - Background Information for Proposed RCRA Air Emission Standards - Volume 2 - Appendices. Preliminary Draft. March 1988, pp. F-146 to F-149.

15. Reference 14, pp. F-144 - F-147.

16. Reference 14, pp. F-151 - F-155.

17. Memorandum from Herndon, D. J., Radian Corporation, to Industrial Wastewater file. Summary of facilities reporting use of steam stripping. May 20, 1988.

18. Environmental Science & Engineering, Inc., and SAIC. Plant No. 4 - Organic Chemicals Best Available Technology Long-Term Field Sampling. Prepared for U. S. Environmental Protection Agency, Office of Water Regulations and Standards. July, 1985.

19. Environmental Science & Engineering, Inc., and SAIC. Plant No. 7 - Organic Chemicals Best Available Technology Long-Term Field Sampling. Prepared for U. S. Environmental Protection agency, Office of Water Regulations and Standards. July, 1985.

20. Environmental Science & Engineering, Inc., and SAIC. Plant No. 7 - Organic Chemicals Best Available Technology Long-Term Field Sampling. Prepared for U. S. Environmental Protection agency, Office of Water Regulations and Standards. July, 1985.

21. Environmental Science & Engineering, Inc., and SAIC. Plant No. 15 - Organic Chemicals Best Available Technology Long-Term Field Sampling. Prepared for U. S. Environmental Protection agency, Office of Water Regulations and Standards. July, 1985.

22. U. S. Environmental Protection Agency. Code of Federal Regulations. Pesticide Chemicals Category Effluent Limitations Guidelines, Pretreatment Standards, and New Source Performance Standards. Title 40, Chapter 1, Subchapter N, Part 455. Washington, D. C. Office of the Federal Register. October 4, 1985.

23. U. S. Environmental Protection Agency. Code of Federal Regulations. Pharmaceutical Manufacturing Point Source Category Effluent Limitations Guidelines, Pretreatment Standards, and New Source Performance Standards. Title 40, Chapter 1, Subchapter N, Part 439. Washington, D. C. Office of the Federal Register. October 27, 1983.

24. Advanced System for Process Engineering (ASPEN). Massachusetts Institute of Technology for the Department of Energy. DOE/MC/16481-(3 vols) 1201, 1202, 1203. 1981.

25. Letter from Plant A to Jack Farmer. (Confidential Section 114 response.) October 1986.

26. Letter from Plant B to Jack Farmer. (Confidential Section 114 response.) November 1986.

27. Reference 25.

28. U. S. Environmental Protection Agency/ORD/IERL. Process Design Manual for Stripping of Organics. Cincinnati, Ohio. EPA 600/2-84-139. August 1984.

29. Reference 25.

30. Reference 22.

31. Reference 22.

32. Office of Air Quality, Planning and Standards. U. S. Environmental Protection Agency. Hazardous Waste Treatment, Storage, and Disposal Facilities (TSDF) - Background Information for Proposed RCRA Air Emission Standards, Preliminary Draft. March 1988.

33. Reference 5.

34. Weber, Walter J. and Edward H. Smith, "Removing dissolved organic contaminants from water," Environmental Science Technology., Vol. 20, No. 10, pp. 970-979, 1986.

35. Office of Air Quality Planning and Standards. U. S. Environmental Protection Agency. Research Triangle Park, North Carolina. Air Stripping of Contaminated Water Sources - Air Emissions and Control. EPA-450/3-87-017 (NTIS PB88-106166), August 1987.

36. Cheremisinoff, Paul N., "Haz Waste Treatment and Recovery Systems". Pollution Engineering. Vol. XX, No. 2, pp. 52 - 61. February, 1988.

37. Memorandum from Elliott, J. A., Radian Corporation, to Industrial Wastewater File, Evaluation of Carbon Adsorption as a Control Technology for Reducing Volatile Organic Compounds (VOC) from Industrial Wastewaters. November 30, 1988.

38. Reference 34.

39. U. S. Environmental Protection Agency. VOC (Volatile Organic Compounds) Emissions from Petroleum Refinery Wastewater Systems - Background Information for Proposed Standards. EPA-450/3-85-001a. February 1985, pp. 4-4 to 4-6.

40. Reference , pp. 4-6 to 4-11.

41. Reference 39, pp. 4-9 to 4-11.

42. Reference 39, pp. 4-12 to 4-14.

43. Reference 39, pp. 4-14 to 4-17.

44. Office of Air Quality Planning and Standards. U. S. Environmental Protection Agency. Control Techniques for Volatile Organic Emissions from Stationary Sources. Draft Report. 3rd Edition. March 1986, pp. 3-1 to 3-83.

45. Office of Air Quality Planning and Standards. U. S. Environmental Protection Agency. Carbon Adsorption for Control of VOC Emissions: Theory and Full Scale system Performance. Draft. Radian Corporation. EPA Contract No. 68-02-4378/20. June 6, 1988.

46. Reference 45.

47. Reference 14, pp. F-132 to F-136.

48. Reference 39, pp. 4-25 to 4-27.

49. Reference 45.

50. Reference 39, p. 4-27 to 4-28.

51. Reference 44.

52. Reference 44.

53. Reference 39, p. 4-27.

54. Reference 39, pp. 4-29 to 4-32.

55. Reference 39, pp. 4-32 to 4-34.

56. Reference 39, p. 4-35.

57. Reference 39, pp. 4-35 to 4-36.

58. Reference 45.

59. Reference 39, p. 4-36.

5. Environmental Impacts of Steam Stripping

The purpose of this chapter is to evaluate the environmental impacts associated with steam stripping. Analysis of the environmental impacts includes an evaluation of the air and water pollution impacts, impacts on waste disposal, and impacts on energy use. An assessment of these impacts is presented for a steam stripper (as described in Chapter 4) applied to each of the three example wastewater streams described in Chapter 3.

5.1 IMPACTS ON VOC EMISSIONS USING A STEAM STRIPPER

Chapter 4 describes the design, operation, and performance of steam stripper systems. Steam stripping is a pretreatment control technique that removes organic compounds from wastewater before the wastewater contacts the ambient air. By effectively removing organic compounds from wastewater, steam stripping reduces the potential for VOC emissions to the air during downstream wastewater collection and treatment and improves water quality. However, the steam stripper system, if not controlled and operated properly, can be a source of VOC emissions. Section 5.1.1 quantitatively presents VOC emission reductions achievable for the example waste streams presented in Chapter 3. Section 5.1.2 presents a qualitative discussion on air toxics, and Section 5.1.3 presents a qualitative discussion of potential VOC emissions that can occur from the steam stripper.

5.1.1 VOC Emissions Reduction

The VOC emission reduction achievable by steam stripping a wastewater stream is based on the stripper design, as discussed in Chapter 4, and the characteristics of the wastewater streams such as flow rate, composition, and organic concentration. Table 5-1 presents a summary of the VOC air emission impacts from the three example waste stream schematics described in Chapter 3. Uncontrolled emissions and emissions after application of the steam stripper system described in Chapter 4 are estimated for the three example waste stream schematics described in Chapter 3 and are included in this table.

TABLE 5-1. SUMMARY OF THE ESTIMATED ANNUAL VOC EMISSION IMPACTS TO THE AIR FROM EACH OF THE EXAMPLE WASTE STREAM SCHEMATICS DEVELOPED IN CHAPTER 3

Flow Rate (lpm)	Example Waste Stream I[a]		Example Waste Stream II[a]		Example Waste Stream III[a]	
	Uncontrolled VOC Emissions (Mg/yr)	Controlled VOC Emissions[b,c] (Mg/yr)	Uncontrolled VOC Emissions (Mg/yr)	Controlled VOC Emissions[b,d] (Mg/yr)	Uncontrolled VOC Emissions (Mg/yr)	Controlled VOC Emissions[b,e] (Mg/yr)
40	19	0.4	24	1.8	42	8.4
150	70	1.5	90	6.6	160	31
300	140	2.9	180	13	320	63
455	210	4.4	270	20	480	95
760	360	7.3	460	33	810	160

[a]The emissions are based on the example wastewater stream described in Table 3-3.
[b]Controlled VOC emissions are those emissions which will occur from the wastewater after steam stripping.
[c]Based on an emission reduction of 98% (See Appendix A, Section A.3) for Example Waste Stream I with application of steam stripping.
[d]Based on an emission reduction of 93% (See Appendix A, Section A.3) for Example Waste Stream II with application of steam stripping.
[e]Based on an emission reduction of 80% (See Appendix A, Section A.3) for Example Waste Stream III with application of steam stripping.

5.1.2 Air Toxics

This guidance document is written for the control of VOC emissions. However, there are other environmental benefits gained from controlling VOC emissions from industrial wastewater. Of particular concern are VOC which are considered hazardous air pollutants. Hazardous air pollutants are addressed in Section 112 of the Clean Air Act (CAA). A compound is currently placed on the intent to list if it is scientifically determined that the compound poses a serious health risk to humans. Standards are then written to regulate emissions of that compound. Amendments to the Clean Air Act are expected to be passed by Congress early in 1990. The amendments call for technology-based rather than health-based standards for hazardous air pollutants. In these amendments, 191 chemicals are listed as hazardous air pollutants. A pollutant is added to the list if it is known to cause or can reasonably be anticipated to cause in humans any of the following: (1) cancer or developmental effects; or (2) serious or irreversible reproductive dysfunctions, neurological disorders, heritable gene mutations, other chronic health effects, or adverse acute human health effects.

Of the 191 listed hazardous air pollutants, approximately 90 percent are VOC. Therefore, for those wastewater streams requiring control of VOC, emissions of hazardous air pollutants could also be significantly reduced.

5.1.3 Steam Stripper Contribution to VOC Emissions

A properly operated and controlled steam stripper can achieve reductions in VOC emissions from industrial wastewaters. However, if improperly designed or operated, the steam stripper may be a source of VOC emissions to the air.

The operating principle of a steam stripper is that steam is contacted with wastewater to provide heat for vaporization of the organic compounds. This produces an overhead vapor that is concentrated in organic compounds from which the organic compounds can be recovered by condensing and decanting.

If the condenser system for the steam stripper does not adequately condense the overhead stream, and is not vented to a control device, the steam stripper may become a concentrated VOC emission source. Other potential emissions from the steam stripper are from the feed tank and decant tank. These tanks would typically have a concentrated organic layer on top of the wastewater and should be vented to a control device (e.g., boiler, process heater, carbon adsorber).

5.2 SECONDARY AIR IMPACTS

This section evaluates the secondary emissions associated with steam stripping. These secondary emissions are then compared to the VOC emission reductions for the three example waste streams.

Secondary air impacts may occur from two sources: combustion of fossil fuels for steam and electricity generation, and handling or combustion of the recovered organics. Fuel combustion for steam and electricity generation is a source of combustion pollutants - particulate matter (PM), sulfur dioxide (SO_2), nitrogen oxides (NO_x), carbon monoxide (CO), and VOC.

The secondary emissions presented in this section are estimated using EPA emission factors which are presented in Table 5-2.[1] These factors assume that steam is generated on site, and electricity is purchased from a local electric utility. Assumptions concerning the fuel composition and boiler efficiencies achieved by the respective generators are based on information compiled by EPA and the Energy Information Administration.[2,3] Adjustments to these values to accommodate emission reductions by existing control devices are made assuming typical controls and control efficiencies presented in these sources.

The industrial boiler used for steam generation is assumed to have a capacity of less than 150 million BTU. An efficiency of 80 percent is assigned to the industrial boiler as an average expected value. It is assumed to be controlled for SO_2, PM, and NO_x emissions using desulfurization (90 percent removal efficiency), an electrostatic precipitator (ESP, 99 percent removal efficiency), and flue gas recirculation (assuming the mid-range of 40 percent removal efficiency), respectively.[4,5] For the purpose of estimating secondary emissions, a fuel composition based on national fuel use for industrial boilers was used. This fuel composition is: natural gas at 45 percent, residual oil at 28 percent, distillate oil at seven percent and coal at 20 percent.[2] Table 5-3 presents a summary of the annual fuel usages for steam and electricity generation. These values are based on the steam stripper design presented in Chapter 4 at the 300 lpm flow rate. The steam requirement, assuming a boiler efficiency of 80 percent, is 3.3×10^{10} kilojoules per year. Average heating values are 2.6×10^{-5} m^3/KJ for natural gas, 2.4×10^{-8} m^3/KJ for residual oil, 2.6×10^{-8} m^3/KJ for distillate oil, and 2.8×10^4 KJ/kg for coal.

TABLE 5-2. COMBUSTION POLLUTANT EMISSION FACTORS

	PM	SO_2	NO_x	CO	VOC
Steam/Electricity Generation					
Natural Gas (kg/10^5 m^3)	4.8	1.0	440	64	2.2
Residual Oil (kg/m^3)[a]	1.6	19.0	6.6	0.60	0.034
Distillate Oil (kg/m^3)[a]	0.24	17.0	2.4	0.60	0.024
Pulverized Coal (g/kg)	60[b]	29.0[c]	11	0.30	0.035

[a]Assumes 1.0% sulfur content in the fuel oil.
[b]Factor derived from the EPA emission factor given as 10A, where A = % ash in coal which was assumed to be a typical value of 12%.
[c]Factor derived from the EPA emission factor given as 39S, where S = % sulfur in coal which was assumed to be the mid-range at 1.5%.

TABLE 5-3. ANNUAL FUEL USE FOR STEAM AND ELECTRICITY GENERATION[a]

Fuel	Percent Composition[b]	Annual Use
Steam Generation		
Natural Gas	45	3.8 x 10^5 sm^3/yr
Residual Oil	28	220 m^3/yr
Distillate Oil	7	59 m^3/yr
Coal	20	2.3 x 10^5 kg/yr
Electricity Generation		
Natural Gas	22	2.0 x 10^3 sm^3/yr
Residual Oil	11	0.92 m^3/yr
Distillate Oil	11	0.99 m^3/yr
Coal	56	6.9 x 10^3 kg/yr

[a]Based on steam stripper design in Chapter 4 at the 300 lpm flow rate case.
[b]Based on national fuel use for industrial and electrical generating boilers.[2]

An electric generator is assumed to use boilers with a higher efficiency (85 percent for this example). The fuel composition, based on national fuel use for electrical generation, is: natural gas at 22 percent, distillate oil at 11 percent, residual oil at 11 percent, and coal at 56 percent. The electrical requirement for the 300 lpm case is 82,820 kWh/yr (see Table 6-3), or 333 MMBtu/yr, based on an 85 percent boiler efficiency. Annual fuel usages are presented in Table 5-3.

Estimated emissions, based on these assumptions are calculated as follows:

uncontrolled emissions = annual fuel use x emission factor, and
controlled emissions = (1 - uncontrolled emissions x control efficiency).

The resulting emission estimates are presented in Table 5-4.

An evaluation of the emission estimates reveal that combustion pollutants associated with electricity generation are small compared to those generated during steam production. In general, combustion pollutant emissions for PM, SO_2, NO_x, CO and VOC are the result of steam needs with a small percentage added, one to ten percent, to account for electricity demand.

Air impacts are presented graphically in Figures 5-1 and 5-2. The values for combustion pollutants are the sum of the contributions from steam and electricity generation.

Handling of the recovered organics may also contribute to secondary air impacts resulting from disposal options. Incineration, for example, produces combustion pollutants. If the recovered organics are recycled, however, they do not contribute to the secondary air impacts. The recovered organics could, in fact, be used as an alternate energy source, i.e., to generate some of the steam required by the stripper. Although combustion of the organics will produce combustion pollutants as mentioned above, emissions of SO_2 and PM would generally be less than those generated by fossil fuel combustion. This is due primarily to two factors: (1) most organic compounds do not contain sulfur, which emits SO_2 when burned; and (2) organic compounds inherently do not contain inorganics, which are emitted as particulates when burned.

TABLE 5-4. SUMMARY OF COMBUSTION POLLUTANT EMISSIONS ASSOCIATED WITH A STEAM STRIPPER

	PM	SO_2	NO_x	CO	VOC
Steam Generation[a,b]					
Natural Gas (Mg/yr)	0.0181	0.00363	1.66	0.242	0.00846
Distillate Oil (Mg/yr)	0.0141	1.00	0.141	0.0353	0.00141
Residual Oil (Mg/yr)	0.342	4.13	1.45	0.132	0.00737
Coal (Mg/yr)	13.8	6.73	2.42	0.0690	0.00805
Subtotal (Mg/yr)	14.2	11.9	5.67	0.478	0.0253
Controlled Emissions (Mg/yr)	0.142	1.19	3.40	0.478	0.0253
Electrical Generation[b,c]					
Natural Gas (Mg/yr)	0.000952	0.000190	0.0873	0.0127	0.000444
Distillate Oil (Mg/yr)	0.00238	0.169	0.0238	0.00595	0.000238
Residual Oil (Mg/yr)	0.0144	0.174	0.0611	0.00555	0.000310
Coal (Mg/yr)	0.415	0.202	0.0727	0.00208	0.000242
Subtotal (Mg/yr)	0.433	0.546	0.245	0.0263	0.00124
Controlled Emissions (Mg/yr)	0.00433	0.0546	0.147	0.0263	0.00124
Total Controlled Emissions (Mg/yr)	0.146	1.24	3.55	0.504	0.0265

[a]Fuel composition for steam generation is based on 45, 28, 7, and 20 percent natural gas, residual oil, distillate oil, and coal, respectively.
[b]SO_2, NO_x, and PM controls reduce emissions by 90, 40, and 99 percent, respectively.[4,5]
[c]Fuel composition for electricity generation is based on 22, 11, 11, and 56 percent natural gas, residual oil, distillate oil, and coal, respectively.

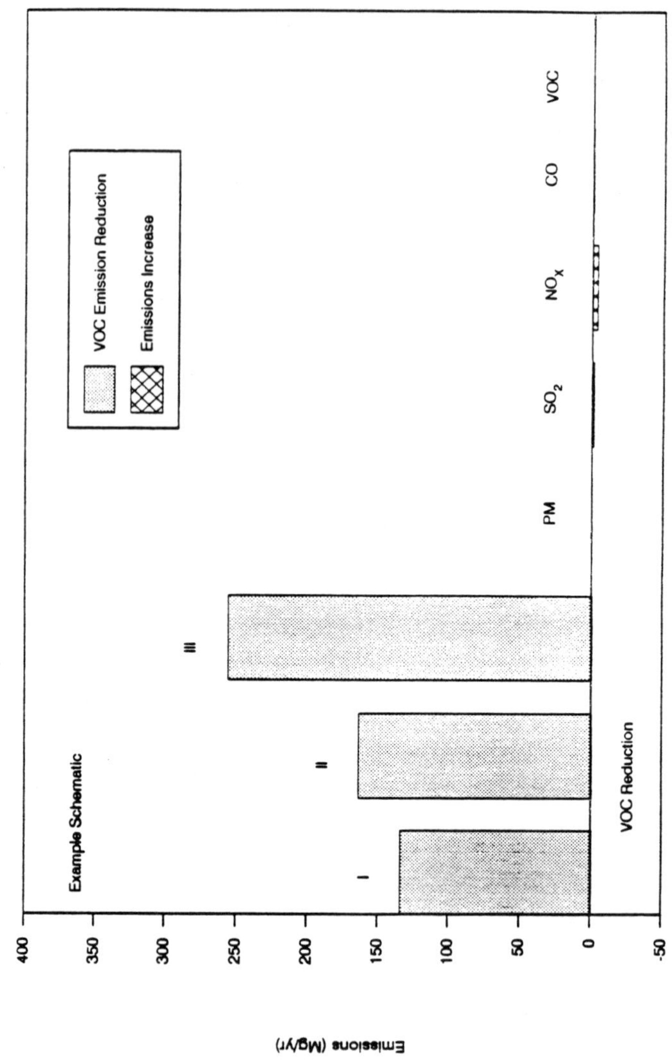

Figure 5-1. Air Impacts of Steam Stripper Control (Controlled Boiler)

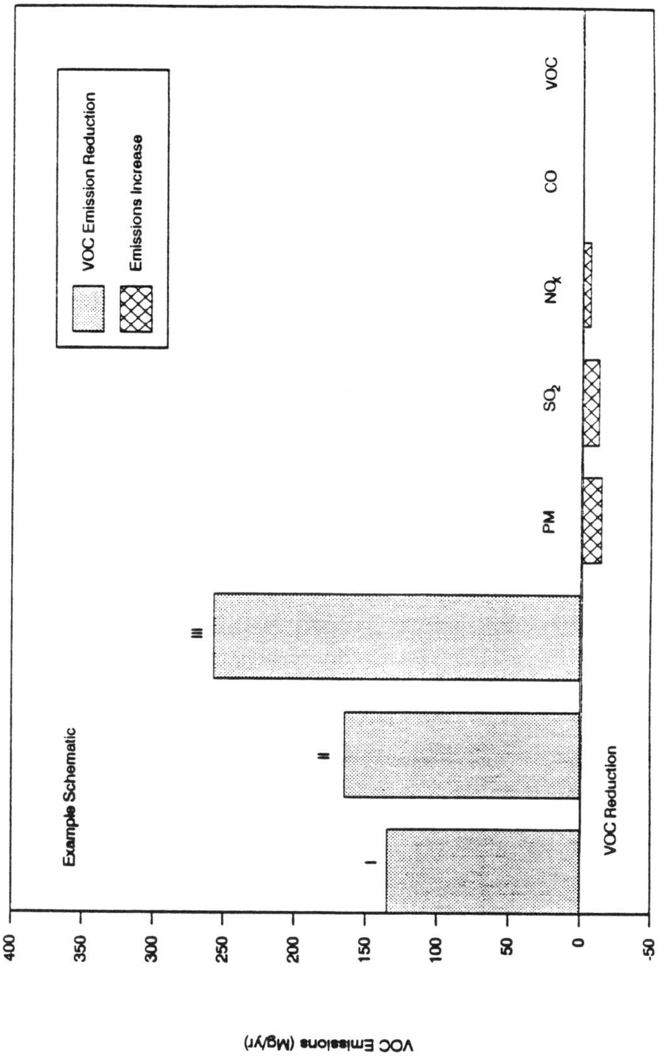

Figure 5-2. Air Impacts of Steam Stripper Control (Uncontrolled Boiler)

5.3 CROSS MEDIA IMPACTS

Other environmental impacts may result from implementation of steam stripper control. A major impact from the use of steam strippers is the improvement in the quality of the wastewater being discharged either directly from the treatment facility or from a POTW into a natural body of water. Also, as mentioned in Section 5.2, organics recovered from the steam stripper can be used as an alternate energy source.

Other impacts include waste disposal, additional demand for non-renewable fuel resources, and the demand of nutrients for biodegradation. Waste generation may arise from three possible sources: disposal of recovered organics, solids removed during feed pretreatment, and control of system vent emissions.

Although an increase in waste generation may occur for non-recyclable organics which cannot be used as supplemental fuel and for cases where treatment is required prior to stripping, it is important to recognize that these organic and/or solid wastes would most likely have been removed from the wastewater anyway (via the air (organics only), an oil/water separator, a clarifier, or activated sludge unit (solids only), for example).

System vent emissions may be sent to a combustion device, thereby generating combustion pollutants, or collected on a sorbent medium that, unless regenerated on site, requires disposal. However, the secondary impacts caused by these types of combustion pollutants are negligible. If sorbent wastes are present, they may adversely impact the soil and/or water depending on whether it is disposed of or regenerated. Therefore, if these disposal methods are necessary, measures should be taken to control emissions from these potential air emission sources.

Another potential impact from steam stripping of organic compounds is the demand of nutrients for biodegradation. If a biobasin is included in the treatment system after the steam stripper, the quantity of nutrients entering the biobasin will be reduced. This may reduce the efficiency of the system.

5.4 ENERGY IMPACTS

The additional fuel demand to generate steam and electricity for the steam stripper system reduces available non-renewable resources: coal, oil, and natural gas. This is partially offset if the recovered organics are used as supplementary fuel or if they are recycled (recycling reduces the facility demand for petroleum-derived feedstocks). It also reduces VOC emissions that could result if the recovered organics had to be disposed of in a waste management facility.

5.5 REFERENCES

1. Office of Air Quality Planning and Standards. U. S. Environmental Protection Agency. Research Triangle Park, North Carolina. Compilation of Air Pollutant Emission Factors. Volume I: Stationary Point and Area Sources. 4th ed., AP-42 (NTIS PB86-124906/REB), September 1985, and AP-42-SUPPL-A (NTIS PB87-150959/REB), October 1986. pp. 1.1-2, 1.3-2, 1.4-2.

2. Office of Air Quality Planning and Standards. U. S. Environmental Protection Agency. Research Triangle Park, North Carolina. Fossil Fuel Fired Industrial Boilers - Background Information. Volume I: Chapters 1-9 Draft EIS. EPA-450/3-82-006a (NTIS PB82-202573) March 1982. p. 3-12-18.

3. Energy Information Administration. U. S. Department of Energy. Washington, D. C. Electric Power Quarterly, April - June 1984. DOE/EIA-0397(84/2Q) October 1984. pp. 19, 20.

4. Reference 1. pp. 1.3-9, 1.3-4.

5. Reference 1. pp. 1.1-5,6.

6. Control Cost Analysis of Steam Stripper System

6.1 STEAM STRIPPER SYSTEM

Steam strippers were discussed in Chapter 4 as an effective emission control strategy. The costs of installing a steam stripper system for removal of organic compounds from industrial wastewater streams for the three example waste stream schematics presented in Chapter 3 are presented in this section. A typical steam stripper system design is presented in Figure 6-1. The wastewater stream is pumped into a storage tank where solids can settle, and some separation of the organic and aqueous phases will occur. Vapors that escape through the pressure/vacuum vent are controlled by an existing combustion or other control device. The wastewater is then pumped to the steam stripper where organics are stripped from the wastewater stream and carried out of the column with the overhead stream. All of the water and most of the organics in the overhead stream are recovered by the condenser unit. The organic phase of the liquid collected from the overhead stream is recovered from the decanter. The aqueous phase is recycled to the wastewater feed stream. Vapors which are not recovered in the condenser are vented to the feed storage tank, which is controlled by an existing combustion or other control device (e.g., boiler, process heater, carbon adsorber). The bottoms from the steam stripper are sent to an on-site plant wastewater treatment facility or to a POTW. In the following sections steam stripper capital costs, annualized costs, and cost effectiveness are presented for the three example waste stream schematics.

6.1.1 Basis For Capital Costs

The total capital investment (TCI) for a steam stripper system includes the basic equipment costs (BEC), all auxiliary equipment costs, and direct and indirect installation costs. The BEC is the sum of the price of each component of the steam stripper system. Total capital investment is composed of the purchased equipment costs (PEC), direct installation costs, and indirect installation costs. The PEC is composed of the BEC, auxiliary piping and equipment costs, instrumentation, freight and sales tax. The BEC is estimated using published engineering cost estimation techniques. The TCI required for a new steam stripper system is calculated as a direct function of

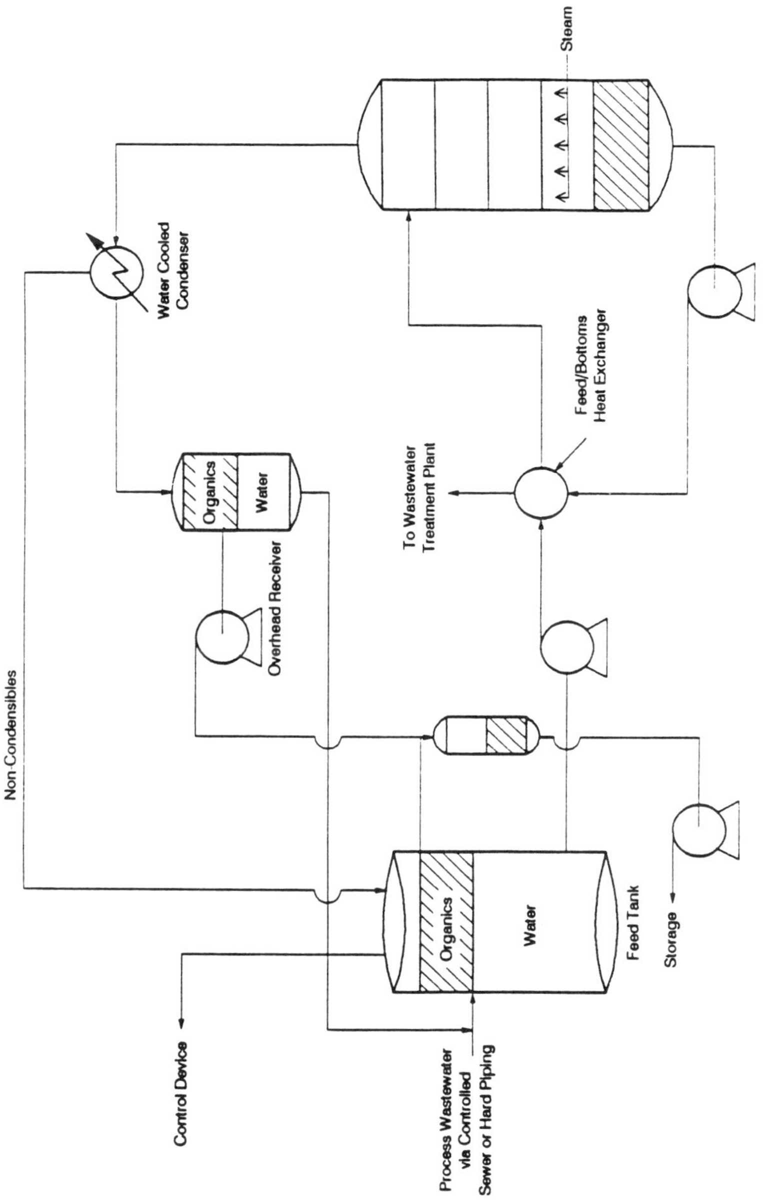

Figure 6-1. Continuous Steam Stripper System

the BEC. These estimation procedures are described more specifically in the following section.

6.1.1.1 Basic Equipment Costs (BEC). Design of the base equipment comprising the steam stripper system shown in Figure 6-1 is based on information gathered by EPA[1,2], and design evaluations performed using Advanced System for Process Engineering (ASPEN)[3], a computer software program for design of distillation columns. Representative wastewater organic concentrations were developed from the information gathered by EPA. The steam stripper system equipment design and operating parameters were then chosen through a design evaluation performed using ASPEN.

The wastewater stream organic concentration and total wastewater throughput vary widely within the target industries. An organic concentration of 2,500 ppm (0.25 percent) at various wastewater throughputs is chosen to represent the wastewater streams for the example schematics. (A sensitivity analysis was performed using ASPEN for a range of organic concentrations. The results showed that removal efficiency varied little within an organic concentration range of 300 to 30,000 ppm.) In addition, a wide variety of organic chemical compounds are present in the wastewater streams. To represent this range, a group of five organic compounds are chosen based on ranges of Henry's Law constants.

The wastewater storage tank is sized to provide a desired retention time of 48 hours for the stripper feed stream, assuming five batch and/or continuous streams are to be combined for treatment by the same steam stripper. Each batch and/or continuous process wastewater stream was assumed to require approximately 300 m of connective piping.

All equipment in the steam stripper system was designed by ASPEN. The steam stripper column is designed as a sieve tray unit with countercurrent flow. The column is operated at a typical steam to wastewater feed ratio of 0.06 kg of steam per liter of wastewater. The liquid loading of the column is 39,900 liters per hour per square meter ($l/hr/m^2$). Based on ASPEN results, an average removal of 67 percent is predicted for the five compounds.

A sensitivity analysis was performed to determine the effect of the column height on the total annualized cost. ASPEN was run at column heights varying from 11.6 to 30.5 meters with all other variables remaining constant. The resultant difference in the ASPEN generated total annualized cost between the shortest and tallest columns was approximately 1.5 percent. Due to the relatively small difference in annual costs, emphasis was placed on generating a design that would be most cost-effective, be within practical design parameters, and would remove virtually 100 percent of the highly volatile compounds. (The controlling compound used for design purposes was benzene.) A column height of 18.3 meters with a total of 31 sieve trays is used for the steam stripper unit.

The overheads from the steam stripper are recovered with a condenser unit consisting of a water-cooled condenser. The condenser is designed for an outlet vapor temperature of 50°C with an overall heat-transfer coefficient (U) of 1,000 joules per square meter per second per degree Kelvin ($j/m^2/s/°K$). The organic phase of the overhead stream is recovered from the overheads decanter. The overhead vapor from the secondary condenser is assumed to be vented to the feed storage tank which is routed to an existing on-site combustion or other control device.

The bottoms from the steam stripper are fed to the existing wastewater treatment facility. Prior to discharge from the stripper system, the bottoms pass through a feed preheater to enhance the efficiency of the steam stripper. The overall heat transfer coefficient used by ASPEN for the feed preheater is 1,000 $j/m^2/s/°K$.

Wastewater is pumped to the feed/bottoms heat exchanger from the stripper to the feed/bottoms heat exchanger (bottoms stream), to the collection pot from the decanter, and from the collection pot to storage.

A total of five vents in the storage tank and decanter are vented to a flare.

Steam stripper costs are estimated using the equipment size generated by ASPEN.[3] The cost of each piece of process equipment is determined from standard chemical engineering cost estimation manuals, textbooks, or journal articles. The cost estimating techniques presented in these sources are based upon the size or capacity of the equipment and are derived from actual construction projects. Table 6-1 summarizes the estimated equipment costs calculated for each component, the estimated size or capacity, and the reference or information source used to obtain the cost estimate for 300 lpm. The initial estimates were based on the equipment costs for the year in which the textbook or journal article was published. These costs were then adjusted to January 1986 dollars using the Chemical Engineering fabricated equipment index for the appropriate month and year. The cost for each individual component was summed to yield the BEC for the example wastewater stream.

TABLE 6-1. ESTIMATION OF BASIC EQUIPMENT COST FOR A STEAM STRIPPING UNIT

Equipment Component	Equipment Size[a]	Construction Material	Equipment Costs[b]	Cost Reference
Feed Tanks	970 m^3	Carbon Steel	\$68,000	4
Feed Preheater (Shell and Tube)	240 m^2	Carbon Steel	\$29,000	4
Steam Stripping Tray Column	0.76 m diameter 18.3 m height	Carbon Steel Trays-Stainless	\$76,000	5,6
Primary Condenser (Water Cooled, Shell & Tube)	22 m^2	Carbon Steel	\$ 5,500	7
Overhead Collection Decanter	2.5 m^3	Carbon Steel	\$ 4,200	8
Flame Arrestor	One arrestor per ventline	NA	\$ 500	9
Pumps (4)	6100 total Watt	Stainless Steel	\$38,000	7
TOTAL BASE EQUIPMENT COST (BEC)			\$220,000	

[a]Based on 300 lpm wastewater flow
[b]January 1986 dollars

NA = Not Applicable

6.1.1.2 Total Capital Investment (TCI).[10] The TCI required to install a new steam stripper unit is calculated as a direct function of the BEC value. The TCI value is composed of three separate components; PEC, direct installation costs, and indirect installation costs. The TCI for the steam stripper unit and the values of each component of the TCI are presented in Table 6-2. The PEC is calculated by multiplying the BEC times an appropriate percentage value. These values are selected from ranges recommended in cost estimation reference documents. Piping costs are implicitly included in the direct installation costs; however, auxiliary piping (i.e., additional piping for the combination of wastewater streams and vapor vent lines for storage tanks) are accounted for separately in the PEC. Instrumentation, sales tax, and freight are also components of the PEC. The PEC is used to estimate the steam stripper system direct installation costs and indirect installation costs. Each of these costs is calculated by multiplying the PEC by an appropriate percentage value. The direct installation costs include items such as electrical wiring, insulation, equipment support and erection, and painting of equipment. The indirect installation costs include engineering, construction and field expense, construction fee, start-up and testing, and contingency. The total of PEC, direct installation costs, and indirect installation costs yields the TCI. The TCI can also include costs for buildings, off-site facilities, land, working capital, and yard improvements; however, these costs are not typically included in the PEC for a steam stripper system.

6.1.1.3 Total Capital Investment versus Wastewater Throughput. The TCI for installing a new steam stripper system is compared to the wastewater throughput in Figure 6-2. TCI is presented for both carbon and stainless steel construction versus flow rates of 40, 150, 300, 455, and 760 lpm. Stainless steel construction costs are included for comparison of streams with corrosive wastewater (i.e., very high or low pH). Equipment costs for stainless steel were developed from the same sources as for carbon steel. Generally, a factor for material of construction was used for conversion of carbon steel to 304 stainless steel. As shown in Figure 6-2, the TCI is a direct function of the wastewater flow rate to the steam stripper unit with stainless steel construction being more costly than carbon steel.

TABLE 6-2. ESTIMATION OF TOTAL CAPITAL INVESTMENT FOR A STEAM STRIPPING UNIT[a,b]

Cost Component	Cost Factor	Component Cost[a]	Total Capital Investment	Cost Reference
Direct Equipment Costs[b]				
Base Equipment Cost (BEC)	TABLE 6-1	$220,000		
Piping[c]	$33.48/m	$53,000		11
Instrumentaton	0.1*[BEC + Pipe]	$27,000		12
Sales Tax and Freight	0.08 * [BEC + Pipe]	$24,000		12
Purchased Equipment Cost (PEC)			$320,000	
Direct Installation Costs				
Foundations and Supports	12% of PEC	$39,000		12
Electrical	1% of PEC	$3,300		12
Erection and Handling	40% of PEC	$130,000		12
Painting	1% of PEC	$3,300		12
Insulation	1% of PEC	$3,300		12
TOTAL DIRECT INSTALLATION COST			$180,000	
Indirect Installation Costs				
Engineering and Supervision	10% of PEC	$33,000		12
Construction & Field Expense	10% of PEC	$33,000		12
Construction Fee	10% of PEC	$33,000		12
Start-up and Testing	1% of PEC	$3,300		12
Contingency	3% of PEC	$9,800		12
TOTAL INDIRECT INSTALLATION COST			$110,000	
TOTAL CAPITAL INVESTMENT (TCI)			$610,000	

[a] January 1986 Dollars
[b] Based on 300 lpm wastewater flow.
[c] Additional piping for combination of five wastewater streams is assumed to total approximately 1500 meters. Vapor vent lines required for storage tanks, and decanters. Each vent line was assumed to be 11 m in length and constructed of 5.1 cm diameter schedule 40 steel pipe.

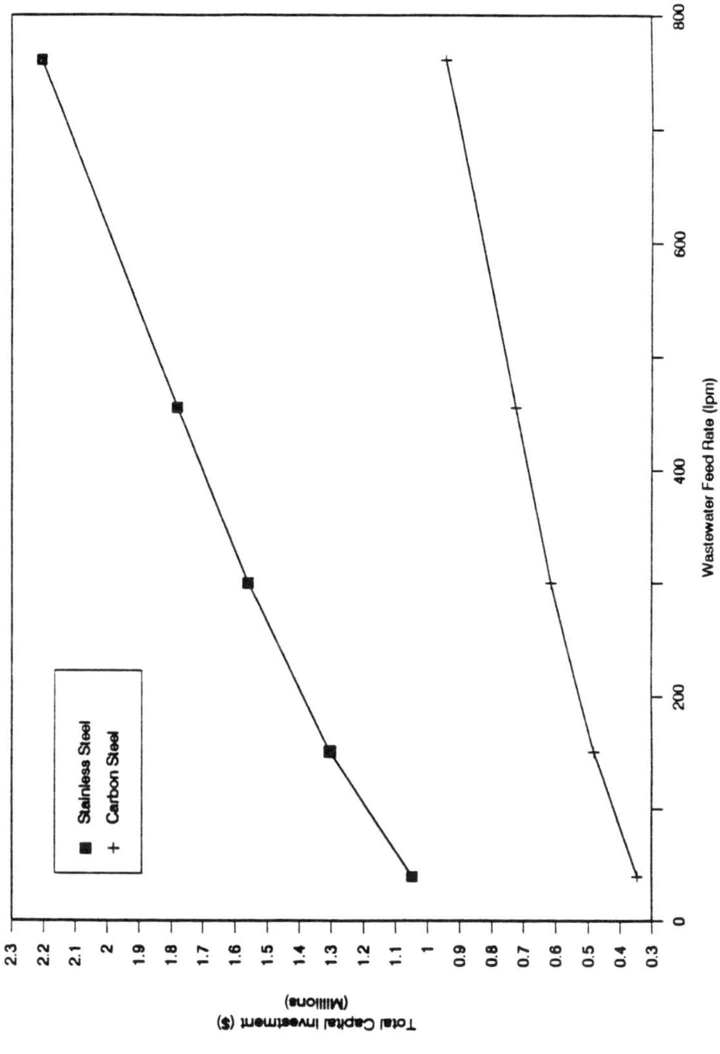

Figure 6-2. Summary of Total Capital Investment versus Wastewater Feed Rate

6.1.2 Basis for Annualized Costs

The total annualized costs (TAC) are the costs incurred to operate the steam stripper process unit throughout the year. The annual operating costs are composed of direct and indirect charges. The TAC and each of its components are presented in Table 6-3.

Direct annual costs are composed of the expenses which are incurred during normal operation of the steam stripper process. These costs include utilities, labor, and maintenance activities. Three types of utilities are required to operate the steam stripper process unit; electricity, steam, and cooling water. Electricity is required to operate pumps and other electrical components included in the system. The electricity required for the pumps is calculated assuming a developed head of approximately 120 feet of water, a pump efficiency of 64 percent, and using design flow rates to each pump. The steam costs are estimated using the design steam loading; 0.06 kg steam per liter of wastewater feed. The cooling water cost is calculated using water requirements necessary for the overhead primary condenser. Other direct costs include labor and maintenance. Labor cost is calculated by multiplying the estimated number of hours required to operate a steam stripper process unit (0.5 hour/shift) times a $12.00/hr labor rate. The supervisory and administrative costs are estimated as 15 percent of operating labor. The maintenance costs are composed of labor and materials. Maintenance labor cost is estimated with 0.5 hours/shift operation times a $13.20/hour labor rate. Maintenance materials cost is 100 percent of maintenance labor cost.

The indirect operating expenses are incurred regardless of the operating status of the steam stripper system. The cost of overhead is estimated to be 60 percent of all labor and maintenance costs. The remaining components of the indirect annual costs are a percentage of the TCI. Property taxes and insurance are both estimated to be one percent of the TCI while administrative charges are estimated to be two percent of the TCI. The capital recovery for the steam stripper system is calculated based on a 15-year equipment life at an interest rate of 10 percent.

Another aspect of the TAC included in this estimate is the recovery credit. This factor accounts for any cost credits which may result from the organics being recovered from the overheads stream. There are several alternatives for handling the recovered organics. If steam is

TABLE 6-3. ESTIMATION OF TOTAL ANNUAL COST FOR A STEAM STRIPPING UNIT[a,b]

Cost Component	Cost Factor	Annual Consumption	Annual Cost	Cost Reference
Direct Annual Costs				
Utilities				
Electricity	\$0.0463/kWhr	44,000 kWhr[c]	\$2,000	13
Steam	\$6.98/Mg	11,000 Mg[d]	\$76,000	13
Water	\$0.053/1,000 liter	470,000,000 liters[e]	\$25,000	14
Labor				
Operating Labor	\$12.00/hr	450 hrs	\$5,400	12
Supervision & Admin	15% of Op. Labor		\$810	12
Maintenance				
Labor	\$13.20/hr	450 hrs	\$5,900	12
Materials	100% of Maint. Labor		\$5,900	12
TOTAL DIRECT ANNUAL COST (TDAC)			\$120,000	
Indirect Annual Costs				
Overhead	60% of All Labor and Materials		\$11,000	12
Property Taxes	1% of TCI		\$6,200	12
Insurance	1% of TCI		\$6,200	12
Administrative Charges	2% of TCI		\$12,000	12
Capital Recovery (CR)	10% @ 15 yrs		\$80,000	
TOTAL INDIRECT ANNUAL COST (TIAC)			\$120,000	
RECOVERY CREDIT (RC)			\$10,000[f]	
TOTAL ANNUAL COST (TAC)	TDAC + TIAC		\$230,000	
ANNUAL WASTE THROUGHPUT (AWT)		130,000 Mg/yr		
COST PER UNIT WASTEWATER (\$/MG)	TAC/AWT		\$1.70	
COST PER LITER WASTEWATER FEED (\$/l)	TAC/FLOW	160,000,000 l/yr	\$0.0014	

[a]January 1986 dollars
[b]Based on 300 lpm wastewater flow
[c]150 kWhr/day, 300 days/yr
[d]37,000 kg/day, 300 days/yr
[e]1,600,000 liters/day, 300 days/yr
[f]Recovery credit based on approximately 28,000 KJ/Kg

produced on-site, the recovered organics can be used as fuel for the existing boiler. The money saved by not having to purchase conventional fuels (i.e., fuel oil or natural gas) is the recovery credit. Another option is to reuse the recovered organics in the manufacturing process. The organics can be recycled directly to the process in some cases. In other cases, the organics must be separated by distillation before reuse. The costs saved in the purchase of raw materials is the recovery credit; however, this may be offset by the cost of distillation for the recovered organics. Another option for the recovered organics is to sell them to a chemical manufacturer who will recover the separate components of the waste organic stream. However, a cost-effective use for the recovered organics may not exist in all cases. In this case, the plant would have to pay for disposal of the collected organics.

Although there are several options available for disposal or use of the recovered organic stream, for this cost estimate it is assumed that the organics can be used as fuel for an existing boiler. A heating value of approximately 28,000 KJ/Kg was calculated for an organic composition developed from Table 2-4.

The organic compounds used to calculate the heating value were chosen based on the highest concentration values of the compounds reported in the 114 responses: acrylonitrile, carbon tetrachloride, ethanol, formaldehyde, styrene, toluene, and triethylamine. The cost of generating steam is reported to be 2 to 3 times more than the fuel cost in Perry's Chemical Engineering handbook.[15] Therefore, to assess a cost savings for burning organics in place of a typical fuel used (i.e., coal, distillate/residual oil, etc.), the typical fuel cost was assumed to be the steam cost divided by 2.5. The resulting fuel cost is $\$1.3 \times 10^{-6}$/KJ. The recovery credit is calculated by multiplying the VOC removal per year by the calculated organic compound heating value and the estimated fuel cost.

The estimated unit annual cost for the steam stripper system at 300 lpm is $1.74/Mg of wastewater treated. The cost per liter of wastewater at this flow rate was estimated to be $0.0014/l. Annual costs were also estimated for four other plant sizes to assess the impact of plant size on annual operating costs. The same estimation techniques used for the steam stripper unit were used to estimate the annual operating costs for steam stripper units treating 40, 150, 455, and 760 lpm of wastewater. The results of these cost estimates are presented graphically in Figure 6-3. Unit operating costs versus flow for both carbon and stainless steel construction

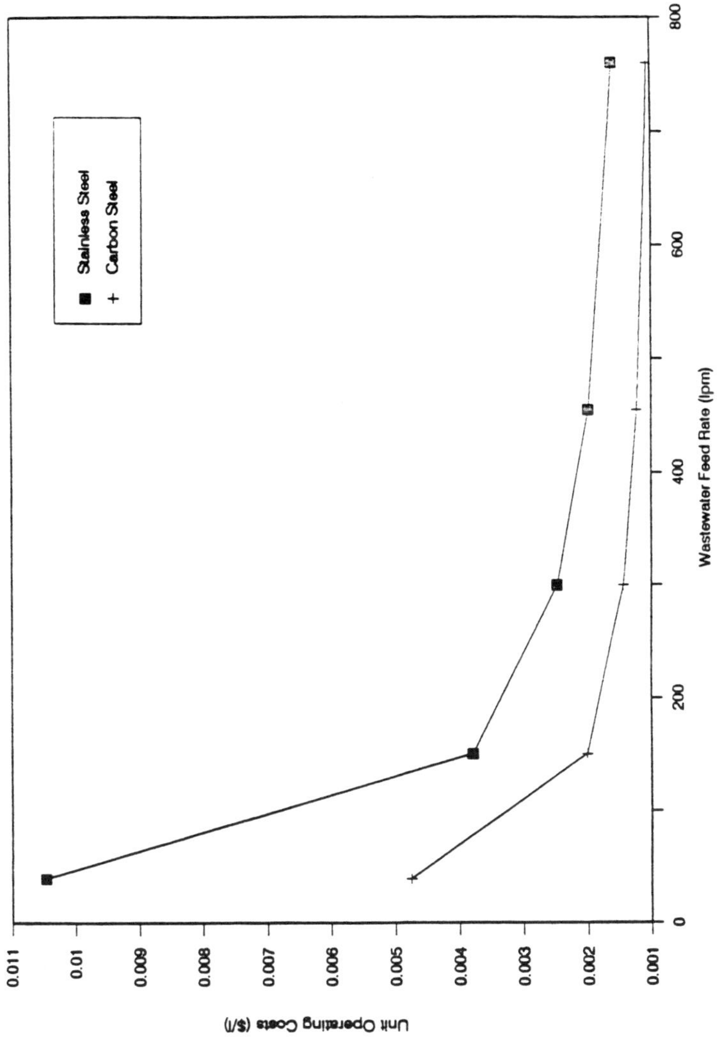

Figure 6-3. Summary of Unit Operating Costs versus Wastewater Feed Rate

are presented. For the carbon steel system, total annual costs (TAC) ranged from \$100,000/yr for the 40 lpm flow rate to \$420,000/yr for a 760 lpm wastewater system. The stainless steel system is more costly with the TAC ranging from \$220,000/yr (40 lpm) to \$640,000/yr (760 lpm).

6.1.3 Cost Effectiveness

The cost effectiveness is defined as the total annualized cost per Mg of VOC emission reduction. The cost effectiveness of steam stripping was calculated for each of the example waste stream collection and treatment system schematics described in Chapter 3 (see Section 3.3). The cost effectiveness was based on a wastewater stream containing 500 ppm (each) of: 1,3-butadiene, toluene, naphthalene, 1-butanol, and phenol. Using the emission factors (fe) generated in Chapter 3, the annual VOC emissions were estimated using the following equation:

$$Q \times 0.001\ m^3/l \times 525{,}600\ min/yr \times 2.5 \times 10^{-3}\ Mg/m^3 \times fe = \text{VOC emissions}$$

where;

Q = wastewater flow rate (lpm)

fe = emission factors generated in Chapter 3 (see Table 6-4).

The VOC emissions from each stream were calculated at five separate wastewater flow rates (Q) ranging from 40 lpm to 760 lpm. The VOC emissions for each of the example wastewater stream schematics at the various flow rates are shown in Table 6-4.

The example waste stream schematics defined in Section 3.3 do not include any type of VOC emission control system. By installing a steam stripper system before wastewaters are exposed to ambient air, VOC emissions could be greatly reduced. Table 6-5 presents controlled and uncontrolled fraction emitted, predicted removal efficiencies for each compound, average organic compound removal efficiency, and VOC emission reduction for each example waste stream schematic applied to the design steam stripper. Table 6-6 presents the estimated annual emissions reduction for the three example waste stream schematics based on the predicted emissions and removal efficiencies. The cost effectiveness for each example waste stream schematic at five different flow rates is presented in Table 6-7 based on the estimated emission

TABLE 6-4. SUMMARY OF THE ESTIMATED ANNUAL VOC EMISSIONS FROM EACH OF THE EXAMPLE WASTE STREAM SCHEMATICS DEVELOPED IN CHAPTER 3

		VOC Emissions (Mg/yr)[a]				
Example Schematic	Emission Factor[b] (fe)	Wastewater Flow Rate (lpm) 40	150	300	455	760
Schematic I	0.35	19	70	140	210	360
Schematic II	0.45	24	90	180	270	460
Schematic III	0.81	42	160	320	480	810

[a]The assumed wastewater organic concentration is 2,500 ppm. (See Table 3-3 for wastewater characteristics).
[b]Overall cumulative fraction emitted for each example schematic. (See Tables A-33 through A-35 in Appendix A.)

TABLE 6-5. REMOVAL EFFICIENCIES AND OVERALL EMISSION REDUCTION

	Uncontrolled Fe[a]				Controlled Fe[c]		
	Example Schematics				Example Schematics		
Compound	I	II	III	Fr[b]	I	II	III
1,3-Butadiene	0.92	1.0	1.00	1.00	0	0	0
Toluene	0.47	0.74	0.997	1.00	0	0	0
Naphthalene	0.34	0.30	0.99	0.999	0.00045	0.00038	0.0013
1-Butanol	0.052	0.22	0.72	0.31	0.036	0.15	0.50
Phenol	0.00079	0.017	0.32	0.022	0.00077	0.016	0.31
Average	0.36	0.45	0.81	0.67	0.0074	0.033	0.16

Emission reduction = [1 - (controlled fe/uncontrolled fe)] * 100

Example Schematic I Emission Reduction (Mg/yr): 0.98
Example Schematic II Emission Reduction (Mg/yr): 0.93
Example Schematic III Emission Reduction (Mg/yr): 0.80

[a]Uncontrolled cumulative fraction emitted for each compound. See Tables A-33 through A-35 of Appendix A.
[b]Fraction removed by the steam stripper is based on ASPEN results for the design steam stripper.[3]
[c]Controlled fraction emitted for each compound. Controlled fe = uncontrolled fe (1-fr).

TABLE 6-6. SUMMARY OF THE ESTIMATED ANNUAL VOC EMISSIONS REDUCTION FROM EACH OF THE EXAMPLE WASTE STREAM SCHEMATICS DEVELOPED IN CHAPTER 3

VOC Emission Reduction (Mg/yr)[a]						
Example Schematic	Emission Reduction[b] (%)	Wastewater Flow Rate (lpm) 40	150	300	455	760
Schematic I	98	18	69	140	210	350
Schematic II	93	22	83	170	250	420
Schematic III	80	34	130	260	390	650

[a]The assumed wastewater organic concentration is 2,500 ppm. (See Table 3-3 for wastewater characteristics).
[b]Determined from ASPEN removal efficiencies and the cumulative uncontrolled fractions emitted for each model plant.[3]

TABLE 6-7. STEAM STRIPPER COST EFFECTIVENESS

	Wastewater Flow Rate (lpm) 40	150	300	455	760
Total Annual Costs (TAC)[b]	$100,000	$160,000	$230,000	$290,000	$420,000
	Cost Effectiveness (dollars/Mg VOC emission reduction)[a]				
Schematic I	5,500	2,300	1,700	1,400	1,200
Schematic II	4,500	1,900	1,400	1,200	1,000
Schematic III	2,900	1,200	890	760	650

[a]The assumed wastewater organic concentration is 2,500 ppm. (See Table 3-3 for wastewater characteristics). Emission reduction was determined from ASPEN removal efficiencies and the cumulative uncontrolled fractions emitted for each example collection and treatment schematic.[3]
[b]Calculated in Section 6.1.2

reductions presented in Table 6-6. Estimates of cost effectiveness are presented graphically in Figure 6-4 for both carbon and stainless steel construction. As shown in Figure 6-4, the cost effectiveness of steam stripping is nearly independent of process throughput at wastewater flow rates greater than 300 lpm. However, at flow rates less than 200 lpm, the cost effectiveness is almost inversely proportional to the total wastewater flow rate. As expected, cost effectiveness for stainless steel construction is greater than that for carbon steel.

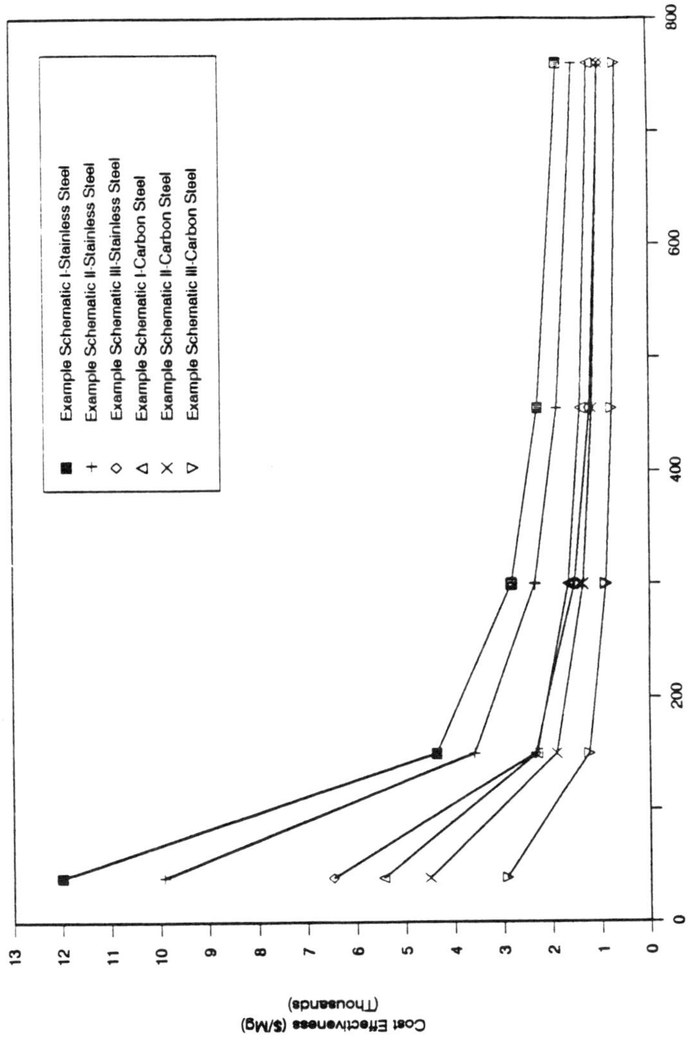

Figure 6-4. Cost Effectiveness versus Wastewater Feed Rate for Example Stream Schematics I, II, III

6.2 REFERENCES

1. Letter from Plant B to Jack Farmer (Confidential Section 114 response.) November 1986.

2. OWRS data, June 12, 1985. JRB/SAIC. "Costing Documentation and National New Information."

3. Advanced System for Process Engineering (ASPEN). Massachusetts Institute of Technology for the Department of Energy. DOE/ME/16481-(3 vols) 1201, 1202, 1203. 1981.

4. Corripio, A. B., K. S. Chrien, and L. B. Evans. Estimate Costs of Heat Exchangers and Storage Tanks via Correlations. "Chemical Engineering", January 25, 1982. p. 125.

5. Peters, M. S., and K. D. Timmerhaus. Plant Design and Economics for Chemical Engineers. 3rd ed. New York, McGraw-Hill Book Company. 1980. pp. 768 - 773.

6. Corripio, A. B., A. Mulet, and L. B. Evans. Estimate Costs of Distillation and Absorption Towers via Correlations. "Chemical Engineering", December 28, 1981. p. 180.

7. Hall, R. S., W. M. Vatavuk, J. Matley. Estimating Process Equipment Costs. "Chemical Engineering", November 21, 1988. p. 66-75.

8. Reference 5, p. 886.

9. Telecon. Gitelman, A., Research Triangle Institute with Hoyt Corporation. September 8, 1986. Cost of flame arrestors.

10. U. S. Environmental Protection Agency. EAB Control Cost Manual. Section 2: Manual Estimating Methodology. 3rd Edition. Draft. Office of Air Quality Planning and Standards. Research Triangle Park, North Carolina. June 1986. p. 2-27 - 2-31.

11. Richardson Process Plant Construction Estimation Standards: Mechanical and Electrical. Volume 3, Richardson Engineering Services, Inc., Mesa, Arizona, 1988. pp 15-40: Pipe costs were adjusted to January 1986 dollars.

12. Vatavuk, W. M., and R. B. Neveril. Part II: Factors for Estimating Capital and Operating Costs. "Chemical Engineering". November 3, 1980. p. 157 - 182.

13. Memorandum from Peterson, Paul, RTI, to Susan Thorneloe, EPA/OAQPS. "Basis for Steam Stripping Organic Removal Efficiency and Cost Estimates Used for the Source Assessment Model (SAM) Analysis". January 18, 1988.

14. Reference 10, p. 4-29.

15. Perry, R. H., and C. H. Chilton. Chemical Engineers' Handbook. 5th ed. New York, McGraw-Hill Book Company. 1973. p. 25-12.

Appendix A—Emission Estimates

Wastewater collection and treatment systems are comprised of a variety of components. A list of the most common components is presented in Table A-1. Because these components are often uncovered and provide contact between wastewater and ambient air, there is the potential for VOC emissions from each of these components. The purpose of this appendix is to present emission estimates for these components and the mass transfer models that were used to perform these estimates. In addition, three example stream schematics were developed by combining individual collection and treatment system components in different scenarios. Cumulative emissions from these three example schematics are also presented in this appendix.

A.1 COLLECTION AND TREATMENT SYSTEM COMPONENTS

Emission estimates for individual collection and treatment system components and the mass transfer models used to estimate these emissions are presented in this section. The emission estimates are based on five example pollutants and typical physical dimensions for the components. The five pollutants are: 1,3-butadiene, toluene, naphthalene, 1-butanol, and phenol. The overall emissions from each component are determined by summing the emissions of these five individual pollutants.

The physical dimensions assigned to each component are based on information gathered by EPA during the Industrial Wastewater project. During this project, 19 chemical manufacturing facilities and 2 pharmaceutical facilities were visited to obtain information on wastewater generation, collection, and treatment.[1-21] During the visits, plant personnel provided data on the dimensions for these components (i.e., the treatment system components listed in Table A-1). Also, during the visits, tours were conducted to evaluate components in the wastewater collection systems. Information gathered during these tours as well as "best engineering judgment" were used to develop "typical" dimensions for these components (i.e., the collection system components listed in Table A-1).

TABLE A-1. WASTEWATER COLLECTION AND TREATMENT SYSTEM COMPONENTS

Collection System Components	Treatment System Components
Drains	Oil-Water Separators
Manholes	Equalization Basins
Junction Boxes	Clarifiers
Sumps	Treatment Tanks
Trenches	Weirs
Lift Stations	Aeration Basins

As discussed in Chapter 3, the emission mechanisms and factors affecting emissions from many of these collection and treatment system components are relatively similar. For this reason, similar mass transfer models are used to estimate emissions from many of the components. The models used to estimate emissions from drains, manholes, and trenches are presented in Section A.1.1. Emissions from junction boxes, lift stations, sumps, equalization basins, clarifiers, treatment tanks, and aeration basins are presented in Section A.1.2 along with the mass transfer models used for these components. Emissions and emission models for oil-water separators and weirs are presented in Section A.1.3 and A.1.4, respectively.

A.1.1 Emissions from Drains, Manholes, and Trenches

As discussed in Chapter 3, the wastewater flowing into process drains and trenches and through underground sewer lines is often greater than the ambient air temperature. For this reason, the air in the headspace at the bottom of the drain riser and in the sewer line may also be greater than the ambient air temperature. Due to this temperature difference, the density of the ambient air is greater than the air in the headspace at the bottom of the drain riser and in the sewer line. Drains, manholes, and trenches provide escape routes for this less dense air to flow into the ambient air above these components. In addition, as wastewater flows through the drain risers, or into trenches it may entrain air flow into the sewer line. This air will escape from the sewer system from downstream drains and manholes or downstream in the trench. Wind blowing into the sewer line or the trench also increases the rate of emissions. Because of these factors, drains, manholes, and trenches are potential emission sources.

Emissions from drains and manholes were estimated based on the factors discussed above at temperatures of 40°C, 30°C, and 30°C, respectively. (Temperatures were chosen based on engineering judgement and the assumption of 25°C ambient temperature.) Typical drain, manhole, and trench configurations are shown in Figure A-1, A-2, and A-3, respectively. Emissions from drains are based on the average of three cases: (1) emissions due to the entrainment of air with the wastewater flowing through the drain into the sewer system, (2) emissions due to wind blowing into the drain, and (3) emissions due to the less dense air flowing from the sewer out the drain riser. Emissions from

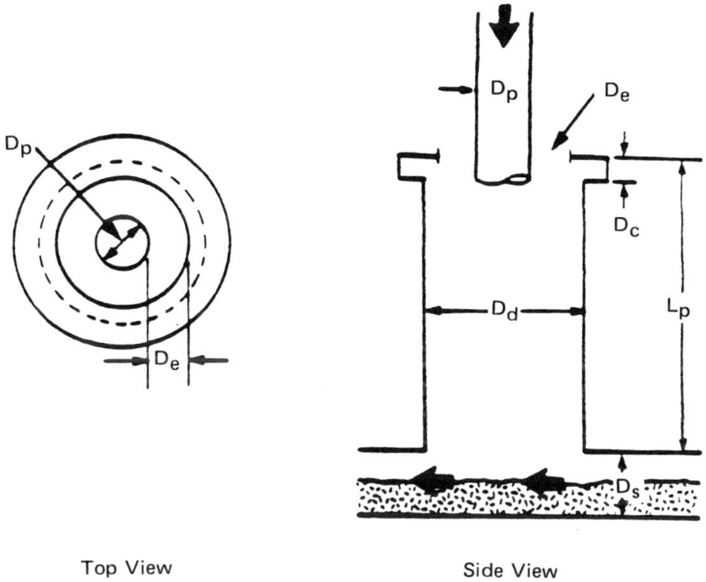

Design Parameter	Range	Typical Design
Drain riser height, L_p (m)	0.3 - 1.2	0.6
Drain riser diameter, D_d (m)	0.1 - 0.3	0.2
Process drain pipe diameter, D_p (m)	0.005 - 0.15	0.1
Effective diameter of drain riser, D_e (m)	0.005 - 0.15	0.1
Drain riser cap thickness, D_c (cm)	0.5 - 0.7	0.6
Sewer Diameter, D_s (m)	0.6 - 1.2	0.9

Figure A-1. Typical drain configuration.

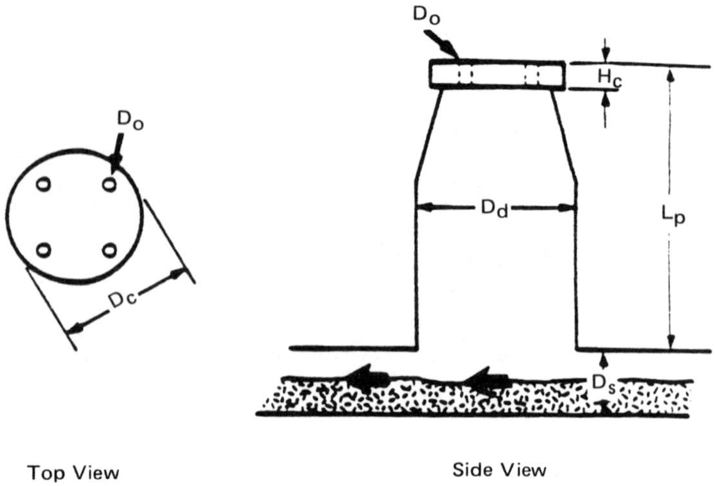

Design Parameter	Range	Typical Design
Manhole diameter, D_d (m)	0.6 - 1.8	1.2
Manhole height, L_p (m)	0.3 - 1.8	1.2
Manhole cover diameter, D_c (m)	0.4 - 0.7	0.6
Diameter of holes in cover, D_o (cm)	1.2 - 3.8	2.5
Manhole cover thickness, H_c (cm)	0.5 - 0.7	0.6
Sewer Diameter, D_s (m)	0.6 - 1.2	0.9

Figure A-2. Typical Manhole Configuration.

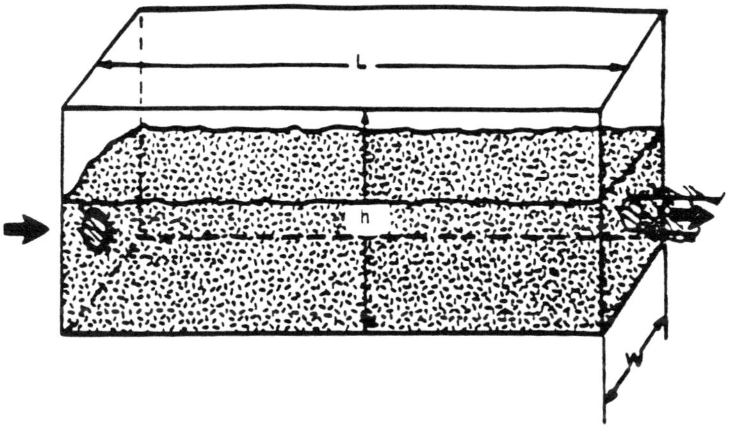

Design Parameter	Range	Typical Design
Trench length, L (m)	15 -150	15.2
Water depth (m)	0.3 - 0.9	0.6
Trench depth, h (m)	0.4 - 1.2	0.8
Trench width, W (m)	0.3 - 0.9	0.6

Figure A-3. Typical trench configuration.

manholes are also based on the average of three cases: (1) emissions due to the less dense air flowing out of the sewer headspace into the ambient air, (2) emissions from wind blowing into the upstream end of a sewer which is obstructed from air flow after the manhole, and 3) emissions from wind blowing through the sewer line under the manhole. Emissions from trenches are based solely on the diffusion of the organics from the flowing wastewater stream. (See Appendix B for the development of emission factors for drains, manholes, and trenches). The emission estimates for drains, manholes, and trenches are presented in Tables A-2, A-3, and A-4, respectively.

A.1.2 Emissions from Junction Boxes, Lift Stations, Sumps, Equalization Basins, Clarifiers, Treatment Tanks, and Aeration Basins

During the Hazardous Waste Treatment, Storage, and Disposal Facilities (TSDF) project, EPA developed mass transfer models to estimate air emissions from surface impoundments and open top tanks.[22] These models are used in this appendix to estimate emissions from junction boxes, lift stations, sumps, equalization basins, clarifiers, treatment tanks, and aeration basins. The models are based on two-film resistance theory. That is, the resistance to mass transfer is assumed to occur at the interface between the liquid (wastewater) and vapor (ambient air) phases. Individual mass transfer coefficients are used to account for these liquid and vapor phase resistances. These individual coefficients are functions of the following parameters: the physical properties of the wastewater pollutants, the design and operation of the collection or treatment system component, and the conditions of the ambient air surrounding the component (e.g., temperature, wind speed). Overall mass transfer coefficients can be determined from the estimated values for the individual coefficients. Then, the general approach used by the models is to apply mass balances around the component and use the overall coefficients to estimate pollutant air emissions. This same general approach is used to estimate emissions from the components discussed in this section. The specific models, assumptions, and component dimensions used to estimate emissions from each component are presented below.

TABLE A-2. EMISSION ESTIMATES FOR DRAINS

Compound	Fraction Emitted, Fe (T=40°C)
1,3-Butadiene	5.7E-01
Toluene	6.1E-02
Naphthalene	1.1E-02
1-Butanol	8.7E-05
Phenol	4.4E-06

TABLE A-3. EMISSION ESTIMATES FOR MANHOLES

Compound	Fraction Emitted, Fe (T=30°C)
1,3-Butadiene	1.5E-01
Toluene	8.2E-03
Naphthalene	1.5E-03
1-Butanol	1.1E-05
Phenol	5.7E-07

TABLE A-4. EMISSION ESTIMATES FOR TRENCHES

Compound	Fraction Emitted, Fe (T = 30°C)
1,3-Butadiene	5.9E-02
Toluene	4.5E-02
Naphthalene	2.5E-02
1-Butanol	4.1E-04
Phenol	2.1E-05

A.1.2.1 Junction Boxes

The general approach used to estimate emissions from junction boxes is presented in Table A-5. Wastewater flowing through the junction box is assumed to be turbulent and well mixed. Correlations are available that can be used to estimate liquid phase coefficients for mechanically aerated impoundments. Although junction boxes are not mechanically aerated, these correlations are used to estimate liquid phase coefficients to account for the wastewater turbulence in the junction box. Air emissions can occur as a result of air entering the sewer system at some point prior to the junction box (i.e., from a manhole or drain). However, it is expected that this type of occurence is rare and the effect on total VOC emissions is small. As seen in Table A-5, the correlation for k_L requires a value for the power input (POWR) to the aerator. Metcalf and Eddy, Inc., suggest a range of 0.5 to 1.0 hp/1,000 ft^3 for mixing in an impoundment.[23] Because the junction box is assumed to be well mixed, a power input of 0.5 hp/1,000 ft^3 is used for this parameter in the correlation to estimate k_L. As shown in Table A-5, individual gas phase mass transfer coefficients are estimated from correlations developed by MacKay and Matasugu. These individual liquid and gas phase coefficients are used to estimate overall mass transfer coefficients for the organic compounds (Step II in Table A-5).

Because the wastewater in the junction box is well mixed, the bulk concentration is equal to the effluent concentration. This bulk concentration provides the driving force for volatilization. Step III in Table A-5 shows the equation for determining the value for this parameter. The overall emission rate for each compound is estimated by determining the product of the following three variables: (1) the liquid surface area in the junction box, (2) the overall mass transfer coefficient for the organic compound, and (3) the concentration of the organic compound in the effluent leaving the junction box (Step IV in Table A-5).

A typical junction box configuration is shown in Figure A-4. Emission estimates for the junction box are based on the typical design dimensions shown in the figure. Table A-6 presents the fraction emitted from the junction box for five example organic compounds. The emission rates presented in the table are based on a wastewater flow rate through the junction box of 150 l/min. The wastewater is assumed to be at a temperature of 35°C. Each of the five organic compounds are assumed to be present in the wastewater feed at

TABLE A-5. ESTIMATION TECHNIQUE FOR JUNCTION BOXES

I. Equations used for calculating liquid and vapor mass transfer coefficients:

<u>Liquid Phase, k_L</u>:

Thibodeaux:

$$k_L\ (m/s) = [8.22 \times 10^{-9}\ J\ (POWR)\ (1.024)^{T-20}\ O_t\ 10^6\ MW_L/(Va_v \rho_L)]\ (D_w/D_{O2,\ w})^{0.5}$$

<u>Vapor Phase</u>, k_G:

Mackay and Matasugu:

$$k_G\ (m/s) = 4.82 \times 10^{-3} \times (U_{10})^{0.78} \times (Sc_G)^{-0.67} \times (d_e)^{-0.11}$$

II. Equation for calculating the overall mass transfer coefficient, K:

$$1/K = 1/k_L + 1/(K_{eq} k_G)$$

where

K_{eq} = Equilibrium constant = H/RT

III. Equation for calculating the bulk concentration of the organic compound:

$$C_L\ (g/m^3) = QC_o/(KA + Q)$$

IV. Equation for calculating air emissions, N_a:

$$N_a\ (Mg/yr) = K\ C_L\ A$$

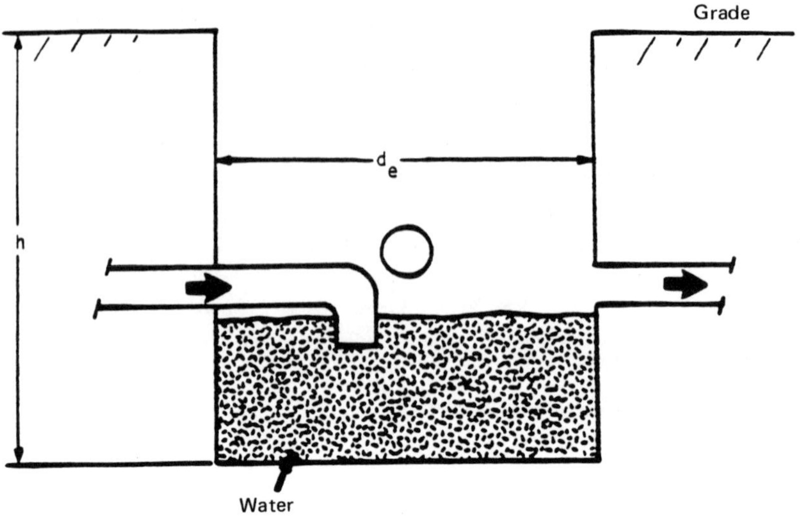

Design Parameter	Range	Typical Design
Effective diameter, d_e (m)	0.3 - 1.8	0.9
Grade height, h (m)	1.2 - 1.8	1.5
Water Depth (m)	0.6 - 1.2	0.9
Surface area (m^2)	0.007 - 2.5	0.7

Figure A-4. Typical Junction Box Configuration.

TABLE A-6. EMISSION ESTIMATES FOR JUNCTION BOXES

Compound	Henry's Law Constant H, atm - m^3/gmol (25°C)	Henry's Law Constant H, atm - m^3/gmol (35°C)	Fraction Emitted, Fe (T=35°C)
1,3-Butadiene	1.42E-01	1.91E-01	1.2E-01
Toluene	6.68E-03	1.10E-02	9.8E-02
Naphthalene	1.18E-03	2.48E-03	6.8E-02
1-Butanol	8.90E-06	1.83E-05	1.9E-03
Phenol	4.54E-07	1.04E-06	1.1E-04

a concentration of 500 ppm. The fraction emitted of each organic compound from the junction box is estimated by dividing the mass of the compound emitted by the mass of that compound in the wastewater. A sample calculation for junction boxes is provided in Table A-7.

A.1.2.2 Lift Stations

The general approach used to estimate emissions from lift stations is presented in Table A-8. Wastewater flowing through the lift station is assumed to be turbulent and well mixed. For this reason, the correlations applicable to mechanically aerated impoundments are used to estimate liquid phase mass transfer coefficients for lift stations. Similar to the procedure used for junction boxes, a power input of 0.5 hp/1,000 ft^3 is used in the correlation to estimate k_L. As shown in Table A-8 (Section I), individual gas phase mass transfer coefficients are estimated from correlations developed by MacKay and Matasugu. These individual liquid and gas phase coefficients are used to estimate overall mass transfer coefficients for the organic compounds in the wastewater (Step II in Table A-8).

Because the wastewater in the lift station is well mixed, the bulk concentration is equal to the effluent concentration. This bulk concentration provides the driving force for volatilization. Step III in Table A-8 shows the equation for determining the value for this parameter. The overall emission rate for each organic compound is estimated by determining the product of the following three variables: (1) the liquid surface area in the lift station, (2) the overall mass transfer coefficient for the organic compound, and (3) the concentration of the organic compound in the effluent leaving the lift station (Step IV in Table A-8).

A typical lift station configuration is shown in Figure A-5. Emission estimates for the lift station are based on the typical design dimensions shown in the figure. Table A-9 presents the fraction emitted from the lift station for five example organic compounds. The emission rates presented in the table based on a wastewater flow rate through the lift station of 150 l/min. The wastewater is assumed to be at a temperature of 30°C. Each of the five organic compounds are assumed to be present in the wastewater feed at a concentration of 500 ppm. The fraction emitted of each compound from the lift station is estimated by dividing the mass of the compound emitted by

TABLE A-7. SAMPLE CALCULATIONS FOR JUNCTION BOXES

Component: Junction Box

$A = 0.656\ m^2$

$d_e = 0.91\ m$

depth = 0.91 m

F/D = 1

$Q = 2.52 \times 10^{-3}\ m^3/s$

t = 236.9 s

T = 35°C

$U_{10} = 4.47\ m/s$

$J = 3\ lb\ O_2/hr\text{-}hp$

$O_t = 0.83$

$N_I = 1$

d = 61 cm

w = 126 rad/s

Fraction agitated = 1

Submerged air flow = $0\ m^3/s$

Compound: Toluene

$C_L = 500\ ppm = 5 \times 10^{-4}\ g/cm^3$

MW = 92.0 g/gmol

$D_w = 1.10 \times 10^{-5}\ cm^2/s$ (35°C)

$D_a = 8.70 \times 10^{-2}\ cm^2/s$

$H = 1.10 \times 10^{-2}\ atm\text{-}m^3/gmol$ (35°C)

$D_{ether} = 8.50 \times 10^{-6}\ cm^2/s$

Water and Air Properties:

$p_w = 1\ g/cm^3$

$u_w = 7.23 \times 10^{-3}\ g/cm\text{-}s$

$p_a = 1.20 \times 10^{-3}\ g/cm^3$

$u_a = 1.81 \times 10^{-4}\ g/cm\text{-}s$

$D_{O2,w} = 3.19 \times 10^{-5}\ cm^2/s$

$MW_a = 29\ g/gmol$

$MW_w = 18\ g/gmol$

TABLE A-7. SAMPLE CALCULATIONS FOR JUNCTION BOXES (Continued)

1. Calculate liquid mass transfer coefficient, k_L:

k_L (m/s) = $[8.22 \times 10^{-9}$ J(POWR) $(1.024)^{T-20}$ $(O_t \times 10^6)$ $(MW_w/(Va_v\rho_w))]$ $[D_w/D_{O2,w}]^{0.5}$

a. Calculate POWR:

POWR = 0.5 hp/1000 $ft^3 \times 35.31\ ft^3/m^3$ (A x depth)

POWR = 0.0005 hp/$ft^3 \times 35.31\ ft^3/m^3$ $(0.656\ m^2 \times 0.91\ m)$

POWR = 0.0105 hp

b. Calculate V:

V = Volume x fraction agitated, ft^3

V = $(7.06\ ft^2 \times 2.986\ ft)$ (1)

V = 21.08 ft^3

c. Calculate a_v:

a_v = area/volume, ft^{-1}

a_v = 7.06 ft/21.08 ft^3

a_v = 0.3349 ft^{-1}

d. Calculate k_L:

k_L (m/s) = $[8.22 \times 10^{-9}$ (3 lb O_2/hr-hp) (0.0105 hp) $(1.024)^{15}]$ $[0.83 \times 10^6$ (18 g/gmol)]/[(21.08 ft^3) (0.3349 ft^{-1}) (1 g/cm^3)] $[1.10 \times 10^{-5}\ cm^2/s\ /\ 3.19 \times 10^{-5}\ cm^2/s]^{0.5}$

k_L (m/s) = (3.696×10^{-10}) (2.116×10^6) (5.872×10^{-1})

k_L = 4.59×10^{-4} m/s

TABLE A-7. SAMPLE CALCULATIONS FOR JUNCTION BOXES (Continued)

2. Calculate vapor mass transfer coefficient, k_g:

k_g (m/s) = $4.82 \times 10^{-3} (U_{10})^{0.78} (Sc_G)^{-0.67} (d_e)^{-0.11}$

a. Calculate Sc_G:

$Sc_G = u_a/p_aD_a$

$Sc_G = 1.81 \times 10^{-4}$ g/cm-s/[(1.20×10^{-3} g/cm^3) (8.70×10^{-2} cm^2/s)]

$Sc_G = 1.734$

b. Calculate k_g:

k_g (m/s) = 4.82×10^{-3} (4.47 m/s)$^{0.78}$ (1.734)$^{-0.67}$ (0.91 m)$^{-0.11}$

k_g (m/s) = 4.82×10^{-3} (3.215) (0.692) (1.0104)

k_g = 0.0108 m/s

3. Calculate overall mass transfer coefficient, K:

$1/K = 1/k_L + 1/(K_{eq}k_g)$

a. Calculate K_{eq}:

$K_{eq} = H/RT$

K_{eq} = 0.0110 atm-m^3/gmol/[(8.21×10^{-5} atm-m^3/gmol-°K) (308°K)]

K_{eq} = 0.434

b. Calculate K:

$1/K = 1/k_L + 1/(K_{eq}k_g)$

$1/K = 1/(4.59 \times 10^{-4}) + 1/[(0.434)(0.0108)]$

$1/K = 2.179 \times 10^3 + 2.133 \times 10^2$

K (m/s) = 4.18×10^{-4}

TABLE A-7. SAMPLE CALCULATIONS FOR JUNCTION BOXES (Continued)

4. Calculate concentration of toluene at vapor-liquid interface, C_L:

$C_L = Q\ C_i/(KA + Q)$

$C_L = (2.52 \times 10^{-3}\ m^3/s)\ (5 \times 10^{-4}\ g/cm^3)\ (10^6\ cm^3/m^3)/$
$[(4.18 \times 10^{-4}\ m/s)\ (0.656\ m^2) + (2.52 \times 10^{-3}\ m^3/s)]$

$C_L = 450.9\ g/m^3$

5. Calculate air emissions, N_a:

$(N_a)_T\ (Mg/yr) = K\ C_L\ A$

$(N_a)_T\ (Mg/yr) = (4.18 \times 10^{-4}\ m/s)\ (3600\ s/hr)\ (7200\ hr/yr)\ (450.9\ g/m^3)$
$(0.656\ m^2)\ (Mg/10^6\ g)$

$(N_a)_T = 3.21\ Mg/yr$

6. Calculate fraction of toluene emitted from a junction box, fe:

$(fe)_T = (N_a)_T/(QC_i)$

$(fe)_T = (3.21\ Mg/yr)/[(2.52 \times 10^{-3}\ m^3/s)\ (3600\ s/hr)\ (7200\ hr/yr)$
$(5 \times 10^{-4}\ g/cm^3)\ (10^6\ cm^3/m^3)\ (Mg/10^6\ g)]$

$(fe)_T = 0.0982$

TABLE A-8. ESTIMATION TECHNIQUE FOR LIFT STATIONS

I. Equations used for calculating liquid and vapor mass transfer coefficients:

Liquid Phase, k_L:

Thibodeaux:

$$k_L\ (m/s) = [8.22 \times 10^{-9}\ J\ (POWR)\ (1.024)^{T-20}\ O_t\ 10^6\ MW_L/(Va_v \rho_L)]\ (D_w/D_{O2,\ w})^{0.5}$$

Vapor Phase, k_G:

Mackay and Matasugu:

$$k_G\ (m/s) = 4.82 \times 10^{-3} \times (U_{10})^{0.78} \times (Sc_G)^{-0.67} \times (d_e)^{-0.11}$$

II. Equation for calculating the overall mass transfer coefficient, K:

$$1/K = 1/k_L + 1/(K_{eq}k_G)$$

where

K_{eq} = Equilibrium constant = H/RT

III. Equation for calculating the bulk concentration of the organic compound:

$$C_L\ (g/m^3) = QC_o/(KA + Q)$$

IV. Equation for calculating air emissions, N_a:

$$N_a\ (Mg/yr) = K\ C_L\ A$$

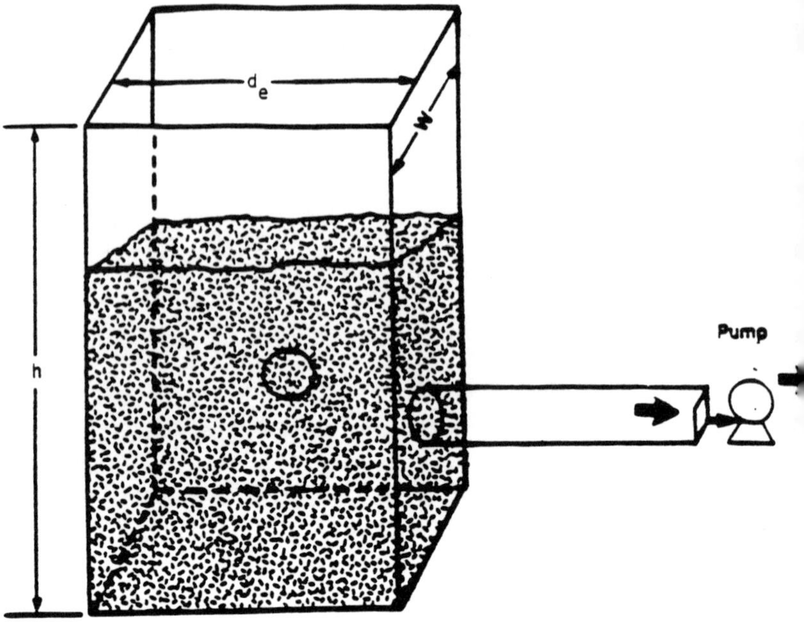

Design Parameter	Range	Typical Design
Effective diameter, d_e (m)	1.2 - 3.0	1.5
Width, W (m)	1.4 - 3.6	1.8
Grade height, h (m)	1.8 - 2.4	2.1
Water Depth (m)	1.2 - 1.8	1.5
Surface area (m^2)	1.1 - 7.1	1.8

Figure A-5. Typical Lift Station Configuration.

TABLE A-9. EMISSION ESTIMATES FOR LIFT STATIONS

Compound	Henry's Law Constant H, atm - m^3/gmol (25°C)	Henry's Law Constant H, atm - m^3/gmol (30°C)	Fraction Emitted, Fe (T=30°C)
1,3-Butadiene	1.42E-01	1.65E-01	3.6E-01
Toluene	6.68E-03	8.61E-03	2.9E-01
Naphthalene	1.18E-03	1.73E-03	1.8E-01
1-Butanol	8.90E-06	1.29E-05	3.6E-03
Phenol	4.54E-07	6.97E-06	2.0E-04

the mass of that compound in the wastewater. An example calculation for lift stations is provided in Table A-10.

A.1.2.3 Sumps

The technique used to estimate emissions from sumps is presented in Table A-11. Individual mass transfer coefficients are estimated based on correlations used for quiescent surface impoundments. Overall mass transfer coefficients are estimated based on values obtained for the individual coefficients. The wastewater flowing through the sump is assumed to be well-mixed. Therefore, the effluent concentration is the driving force for air emissions. The overall coefficients, the liquid surface area in the clarifier, and the wastewater effluent concentrations for each organic compound are then multiplied together to estimate the emission rate of each organic compound.

Table A-12 presents the fraction emitted for five example organic compounds. The emission rates are based on a wastewater flow rate of 150 lpm through the sump. Each of the five organic compounds are assumed to be present at a concentration of 100 ppm. The wastewater is assumed to be at a temperature of 30°C. An example calculation for sumps is shown in Table A-13.

A.1.2.4 Equalization Basins

The technique used to estimate emissions from equalization basins is presented in Table A-14. Equalization basins may be non-aerated or aerated. Emission estimates for these cases are shown in Steps (1) and (2) in the Table A-14. The major difference in the two procedures is the calculation of the individual mass transfer coefficients. Individual mass transfer coefficients for the non-aerated case are estimated based on correlations used for quiescent surface impoundments. Individual mass transfer coefficients for the aerated case are estimated based on correlations used for mechanically aerated surface impoundments. Overall mass transfer coefficients (in both cases) are estimated based on values obtained for the individual coefficients. The overall coefficients, the liquid surface area in the equalization basin, and the wastewater effluent concentrations for each organic compound are then multiplied together to estimate the emission rate of each organic compound.

TABLE A-10. SAMPLE CALCULATIONS FOR LIFT STATIONS

Component: Lift Station

$A = 1.824\ m^2$	$J = 3\ lb\ O_2/hr\text{-}hp$
$d_e = 1.524\ m$	$O_t = 0.83$
depth = 1.524 m	$N_I = 1$
F/D = 1	d = 61 cm
$Q = 2.52 \times 10^{-3}\ m^3/s$	w = 126 rad/s
t = 1,103 s	Fraction agitated = 1
T = 30°C	Submerged air flow = 0 m^3/s
$U_{10} = 4.47\ m/s$	

Compound: Toluene

$C_L = 500\ ppm = 5 \times 10^{-4}\ g/cm_3$

MW = 92.0 g/gmol

$D_w = 9.74 \times 10^{-6}\ cm^2/s$ (30°C)

$D_a = 8.7 \times 10^{-2}\ cm^2/s$

$H = 8.61 \times 10^{-3}\ atm\text{-}m^3/gmol$ (30°C)

Water and Air Properties:

$p_w = 1\ g/cm^3$

$u_w = 8.01 \times 10^{-3}\ g/cm\text{-}s$

$p_a = 1.20 \times 10^{-3}\ g/cm^3$

$u_a = 1.81 \times 10^{-4}\ g/cm\text{-}s$

$D_{O2,w} = 2.83 \times 10^{-5}\ cm^2/s$

$MW_a = 29\ g/gmol$

$MW_w = 18\ g/gmol$

TABLE A-10. SAMPLE CALCULATIONS FOR LIFT STATIONS
(Continued)

1. Calculate liquid mass transfer coefficient, k_L:

$$k_L\ (m/s) = [8.22 \times 10^{-9}\ J(POWR)\ (1.024)^{T-20}\ (O_t \times 10^6)\ (MW_w/(Va_v\rho_w))]\ [D_w/D_{O2,w}]^{0.5}$$

a. Calculate POWR:

$$POWR = 0.5\ hp/1000\ ft^3 \times 35.31\ ft^3/m^3\ (A \times depth)$$

$$POWR = 0.0005\ hp/ft^3 \times 35.31\ ft^3/m^3 \times (1.824\ m^2 \times 1.524\ m)$$

$$POWR = 0.049\ hp$$

b. Calculate V:

$$V = Volume \times fraction\ agitated,\ ft^3$$

$$V = (19.6\ ft^2 \times 5\ ft)\ (1)$$

$$V = 98\ ft^3$$

c. Calculate a_v:

$$a_v = area/volume,\ ft^{-1}$$

$$a_v = 19.6\ ft^2/98\ ft^3$$

$$a_v = 0.2\ ft^{-1}$$

d. Calculate k_L:

$$k_L\ (m/s) = [8.22 \times 10^{-9}\ (3\ lb\ O_2/hr\text{-}hp)\ (0.049\ hp)\ (1.024)^{10}]\ [0.83 \times 10^6\ (18\ g/gmol)/(98\ ft^3)\ (0.2)\ (1\ g/cm^3)]\ [9.74 \times 10^{-6}\ cm^2/s\ /\ 2.83 \times 10^{-5}\ cm^2/s]^{0.5}$$

$$k_L\ (m/s) = (1.53 \times 10^{-9})\ (762245)\ (.586)$$

$$k_L = 6.83 \times 10^{-4}\ m/s$$

TABLE A-10. SAMPLE CALCULATIONS FOR LIFT STATIONS (Continued)

2. Calculate vapor mass transfer coefficient, k_g:

k_g (m/s) = $4.82 \times 10^{-3} (U_{10})^{0.78} (Sc_G)^{-0.67} (d_e)^{-0.11}$

a. Calculate Sc_G:

$Sc_G = u_a/(p_a D_a)$

$Sc_G = 1.81 \times 10^{-4}$ g/cm-s/[(1.20×10^{-3} g/cm^3) (8.70×10^{-2} cm^2/s)]

$Sc_G = 1.734$

b. Calculate k_g:

k_g (m/s) = 4.82×10^{-3} (4.47 m/s)$^{0.78}$ (1.734)$^{-0.67}$ (1.524 m)$^{-0.11}$

k_g (m/s) = 4.82×10^{-3} (3.215) (0.6916) (0.9547)

k_g = 0.0102 m/s

3. Calculate overall mass transfer coefficient, K:

$1/K = 1/k_L + 1/(K_{eq}k_g)$

a. Calculate K_{eq}:

$K_{eq} = H/RT$

K_{eq} = 0.00861 atm-m^3/gmol/[(8.21×10^{-5} atm-m^3/gmol-°K) (303°K)]

K_{eq} = 0.346

b. Calculate K:

$1/K = 1/k_L + 1/(K_{eq}k_g)$

$1/K = 1/6.83 \times 10^{-4} + 1/[(0.346)(0.0102)]$

1/K = 1464.1 + 283.3

K (m/s) = 5.73 x 10-4

TABLE A-10. SAMPLE CALCULATIONS FOR LIFT STATIONS
(Continued)

4. Calculate concentration of toluene at vapor-liquid interface, C_L:

$C_L = Q\ C_i/(KA + Q)$

$C_L = (2.52 \times 10^{-3}\ m^3/s)\ (5 \times 10^{-4}\ g/cm^3)\ (10^6\ cm^3/m^3)/$

$[(5.73 \times 10^{-4}\ m/s)\ (1.824\ m^2) + (2.52 \times 10^{-3}\ m^3/s)]$

$C_L = 353.4\ g/m^3$

5. Calculate air emissions, N_a:

$(N_a)_T\ (Mg/yr) = K\ C_L\ A$

$(N_a)_T\ (Mg/yr) = (5.73 \times 10^{-4}\ m/s)\ (3600\ s/hr)\ (7200\ hr/yr)\ (353.4\ g/m^3)$

$(1.824\ m2)\ (Mg/10^6\ g)$

$(N_a)_T = 9.57\ Mg/yr$

6. Calculate fraction of toluene emitted from a lift station, fe:

$(fe)_T = (N_a)_T/(QC_i)$

$(fe)_T = (9.57\ Mg/yr)/[(2.52 \times 10^{-3}\ m^3/s)\ (3600\ s/hr)\ (7200\ hr/yr)$

$(5 \times 10^{-4}\ g/cm^3)\ (106\ cm^3/m^3)\ (Mg/10^6\ g)]$

$(fe)_T = 0.293$

TABLE A-11. ESTIMATION TECHNIQUE FOR SUMPS

I. Equations used for calculating liquid and vapor mass transfer coefficients:

Liquid Phase, k_L:

Mackay and Yeun
(for F/D <14 and U_{10} >3.25 m/s):

$$k_L = 1.0 \times 10^{-6} + 144 \times 10^{-4} \times (U^*)^{2.2} \times (Sc_L)^{-0.5},$$

where

$$U^* < 0.3$$

where

$$U^* \ (m/s) = 0.01 \times U_{10} \times (6.1 + 0.63 \times U_{10})^{0.5}$$

and

$$Sc_L = u_L/(p_L D_w)$$

Vapor Phase, k_G:

Mackay and Matasugu:

$$k_G \ (m/s) = 4.82 \times 10^{-3} \times (U_{10})^{0.78} \times (Sc_G)^{-0.67} \times (d_e)^{-0.11}$$

II. Equation for calculating the overall mass transfer coefficient, K:

$$1/K = 1/k_L + 1/(K_{eq} k_G)$$

where

$$K_{eq} = \text{Equilibrium constant} = H/RT$$

III. Equation for calculating the bulk concentration of the organic compound:

$$C_L \ (g/m^3) = QC_o/(KA + Q)$$

IV. Equation for calculating air emissions, N_a:

$$N_a \ (Mg/yr) = K \ C_L \ A$$

TABLE A-12. EMISSION ESTIMATES FOR SUMPS

Compound	Henry's Law Constant H, atm - m^3/gmol (25°C)	Henry's Law Constant H, atm - m^3/gmol (30°C)	Fraction Emitted, Fe (T=30°C)
1,3-Butadiene	1.42E-01	1.65E-01	5.6E-03
Toluene	6.68E-03	8.61E-03	5.0E-03
Naphthalene	1.18E-03	1.73E-03	4.7E-03
1-Butanol	8.90E-06	1.29E-05	2.1E-03
Phenol	4.54E-07	6.97E-07	1.9E-04

TABLE A-13. SAMPLE CALCULATIONS FOR SUMPS

Component: Sump

$A = 1.824\ m^2$

$d_e = 1.524\ m$

depth = 1.524 m

F/D = 1.0

$Q = 2.52 \times 10^{-3}\ m^3/s$

t = 1,103 s

T = 30°C

$U_{10} = 4.47\ m/s$

Compound: Toluene

$C_L = 100\ ppm = 1 \times 10^{-4}\ g/cm^3$

MW = 92.0 g/gmol

$D_w = 9.74 \times 10^{-6}\ cm^2/s$

$D_a = 8.70 \times 10^{-2}\ cm^2/s$

$H = 8.61 \times 10^{-3}\ atm\text{-}m^3/gmol$

$D_{ether} = 8.5 \times 10^{-6}\ cm^2/s$

Water and Air Properties:

$p_w = 1\ g/cm^3$

$u_w = 8.01 \times 10^{-3}\ g/cm\text{-}sec$

$p_a = 1.20 \times 10^{-3}\ g/cm^3$

$u_a = 1.81 \times 10^{-4}\ g/cm\text{-}s$

TABLE A-13. SAMPLE CALCULATIONS FOR SUMPS (Continued)

1. Calculate liquid mass transfer coefficient, k_L:

a. Calculate U*:

$U^* \text{ (m/s)} = 0.01 \times U_{10} (6.1 + 0.63 \times U_{10})^{0.5}$

$U^* \text{ (m/s)} = 0.01 \times 4.47 \text{ m/s} (6.1 + 0.63 \times 4.47 \text{ m/s})^{0.5}$

$U^* = 0.1335 \text{ m/s}$

b. Calculate Sc_L:

$Sc_L = u_w/(p_w D_w(T))$

$Sc_L = 8.01 \times 10^{-3} \text{ g/cm-s}/[(1 \text{ g/cm}^3) (9.74 \times 10^{-6} \text{ cm}^2/\text{s})]$

$Sc_L = 822$

c. Calculate k_L:

$k_L \text{ (m/s)} = 1.0 \times 10^{-6} + 144 \times 10^{-4} (U^*)2.2 (Sc_L)^{-0.5}$

$k_L \text{ (m/s)} = 1.0 \times 10^{-6} + 144 \times 10^{-4} (0.1335)^{2.2} \times (822)^{-0.5}$

$k_L = 6.98 \times 10^{-6} \text{ m/s}$

2. Calculate vapor mass transfer coefficient, k_g:

$k_g \text{ (m/s)} = 4.82 \times 10^{-3} \times (U_{10})^{0.78} (Sc_G)^{-0.67} (d_e)^{-0.11}$

a. Calculate Sc_G:

$Sc_G = u_a/(p_a D_a(T))$

$Sc_G = (1.81 \times 10^{-4} \text{ g/cm-s})/[(1.20 \times 10^{-3} \text{ g/cm}^3) (8.70 \times 10^{-2} \text{ cm}^2/\text{s})]$

$Sc_G = 1.734$

b. Calculate k_g:

$k_g \text{ (m/s)} = 4.82 \times 10^{-3} \times (4.47 \text{ m/s})^{0.78} (1.734)^{-0.67} (1.524 \text{ m})^{-0.11}$

$k_g \text{ (m/s)} = 4.82 \times 10^{-3} \times (3.215) (0.692) (0.955)$

$k_g = 0.0102 \text{ m/s}$

TABLE A-13. SAMPLE CALCULATIONS FOR SUMPS (Continued)

3. Calculate overall mass transfer coefficient, K:

$1/K = 1/K_L + 1/(K_{eq}k_g)$

a. Calculate K_{eq}:

$K_{eq} = H/RT$

$K_{eq} = 8.61 \times 10^{-3}$ atm-m^3/gmol/[(8.21 $\times 10^{-5}$ atm-m^3/gmol-°K) (303°K)

$K_{eq} = 0.346$

b. Calculate K:

$1/K = 1/k_L + 1/(K_{eq}k_g)$

$1/K = (1/6.98 \times 10^{-6}$ m/s) + [1/(0.346) (0.0102 m/s)]

$1/K = 143266 + 283.4$

$K = 6.96 \times 10^{-6}$ m/s

4. Calculate concentration of toluene at vapor-liquid interface, C_L:

$C_L = Q\ C_i/(KA + Q)$

$C_L = (2.52 \times 10^{-3}$ m^3/s) (1 $\times 10^{-4}$ g/cm^3) (10^6 cm^3/m^3)/

[(6.96 $\times 10^{-6}$ m/s) (1.824 m^2) + (2.52 $\times 10^{-3}$ m^3/s)

$C_L = 99.5$ g/m^3

5. Calculate air emissions, N_a:

$(N_a)_T$ (Mg/yr) = K C_L A

$(N_a)_T$ (Mg/yr) = (6.96 $\times 10^{-6}$ m/s) (3600 s/hr) (7200 hr/yr)

(99.5 g/m^3) (1.824 m^2) (10^{-6} Mg/g)

$(N_a)_T = 0.0328$ Mg/yr

6. Calculate fraction of toluene emitted from a sump, f_e:

$(f_e)_T = (N_a)_T/QC_i$

$(f_e)_T$ = (0.0328 Mg/yr)/[(2.52 $\times 10^{-3}$ m^3/s) (3600 s/hr) (7200 hr/yr)

(1 $\times 10^{-4}$ g/cm^3) (106cm^3/m^3) (10^{-6} Mg/g)

$(f_e)_T = 5.02 \times 10^{-3}$

TABLE A-14. ESTIMATION TECHNIQUE FOR EQUALIZATION BASINS

(1) Non-Aerated Basins

I. Equations used for calculating liquid and vapor mass transfer coefficients:

Liquid Phase, k_L:

Springer et al (for 14 <F/D <51.2 and U_{10} >3.25 m/s):

$$k_L \text{ (m/s)} = [2.605 \times 10^{-9} (F/D) + 1.277 \times 10^{-7}] \times (U_{10})^2 \times (D_w/D_{ether})^{2/3}$$

Vapor Phase, k_G:

MacKay and Matasugu:

$$k_G \text{ (m/s)} = 4.82 \times 10^{-3} \times (U_{10})^{0.78} \times (Sc_G)^{-0.67} \times (d_e)^{-0.11}$$

II. Equation for calculating the overall mass transfer coefficient, K:

$$1/K = 1/k_L + 1/(K_{eq}k_G)$$

where

$$K_{eq} = \text{Equilibrium constant} = H/RT$$

III. Equation for calculating the bulk concentration of the organic compound:

$$C_L \text{ (g/m}^3) = QC_o/(KA + Q)$$

IV. Equation for calculating air emissions, N_a:

$$N_a \text{ (Mg/yr)} = K\ C_L\ A$$

(2) Aerated Basin

I. Equations used for calculating liquid and vapor mass transfer coefficients:

Liquid Phase, k_L:

$$k_L \text{ (m/s)} = [8.22 \times 10^{-9} J\ (POWR)(1.024)^{T-20}\ O_t\ 10^6\ MW_L/(Va_v p_L)] \\ (D_w/D_{O2,w})^{0.5}$$

TABLE A-14. ESTIMATION TECHNIQUE FOR EQUALIZATION BASINS (Continued)

Gas Phase

Reinhardt:

k_G (m/s) = 1.35 x 10-3-7 $(Re)^{1.42}$ $(P)^{0.4}$ $(Sc_G)^{0.5}$ $(Fr)^{-0.21}$ $(D_a MW_a/d)$

II. Equation for calculating the overall mass transfer coefficient, K;

$$1/K = 1/(k_L) + 1/(K_{eq} k_G)$$

where

K_{eq} = Equilibrium constant = H/RT

Equation for calculating overall mass transfer coefficient for combined quiescent and turbulent areas:

$$K = K_Q A_Q + K_T A_T)/(A_Q + A_T)$$

where

K_Q = overall mass transfer coefficient for quiescent area,

K_T = overall mass transfer coefficient for turbulent area,

A_Q = quiescent surface area, and

A_T = turbulent surface area.

III. Equation for calculating the bulk concentration of the organic compound:

$$C_L\ (g/m^3) = QC_o/(KA + Q)$$

IV. Equation for calculating air emissions, N_a:

$$N_a\ (Mg/yr) = K\ C_L\ A$$

Figure A-6 presents a typical equalization basin. Emission estimates for the equalization basin are based on the typical design dimensions shown in the figure. Table A-15 presents emission estimates for typical aerated and non-aerated equalization basin designs. The emission rates were based on a wastewater flow rate of 1,000 gpm through the equalization basin. Each of the five organic compounds were assumed to be present at a concentration of 10 ppm. The wastewater is assumed to be at ambient temperature for the purposes of the calculation. An example calculation for equalization basins is shown in Table A-16.

A.1.2.5 Clarifiers

The technique used to estimate emissions from clarifiers is presented in Table A-17. Individual mass transfer coefficients are estimated based on correlations used for quiescent surface impoundments. Overall mass transfer coefficients are estimated based on values obtained for the individual coefficients. The wastewater flowing through the clarifier is assumed to be well-mixed. Therefore, the effluent concentration is the driving force for air emissions. The overall coefficients, the liquid surface area in the clarifier, and the wastewater effluent concentrations for each organic compound are then multiplied together to estimate the emission rate of each organic compound.

A typical clarifier is shown in Figure A-7.[24] Emission estimates for the clarifier are based on the typical design dimensions shown in the figure. Table A-18 presents the fraction emitted for five example organic compounds. The emission rates are based on a wastewater flow rate of 1,000 gpm through the clarifier. Each of the five organic compounds are assumed to be present at a concentration of 10 ppm. The wastewater is assumed to be at ambient temperature for the purposes of the calculation. An example calculation for clarifiers is shown in Table A-19.

A.1.2.6 Emissions from Aerated and Non-Aerated Biological Treatment Basins

Mass transfer correlations used during the TSDF project to estimate emissions from activated sludge units and disposal impoundments with quiescent surfaces are used to estimate emissions from aerated and non-aerated biological treatment basins, respectively. These techniques are presented in

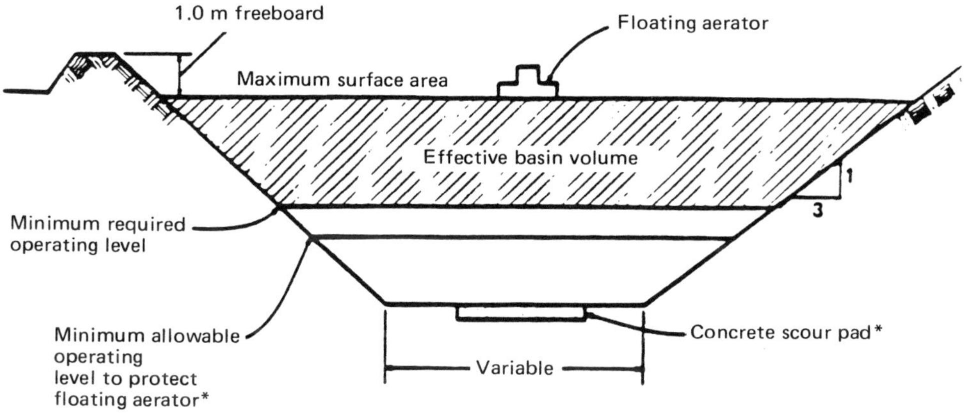

*These dimensions will vary with aerator design and size.

Design Parameter	Range	Typical Design
Effective Diameter (m)	20 -270	109
Surface Area (m^2)	300 - 57,000	9,290
Water depth (m)	1 - 8	2.9
Retention time (days)	0.2 - 20	5

Figure A-6. Typical equalization basin.

TABLE A-15. EMISSION ESTIMATES FOR EQUALIZATION BASINS

Compound	Henry's Law Constant H, atm - m^3/gmol (25°C)	Fraction Emitted, Fe
Non-Aerated		
1,3 Butadiene	1.42E-01	4.3E-01
Toluene	6.68E-03	4.0E-01
Naphthalene	1.18E-03	3.8E-01
1-Butanol	8.90E-06	1.8E-01
Phenol	4.54E-07	1.6E-02
Aerated		
1,3 Butadiene	1.42E-01	1.0E+00
Toluene	6.68E-03	9.9E-01
Naphthalene	1.18E-03	9.88E-01
1-Butanol	8.90E-06	6.1E-01
Phenol	4.54E-07	7.7E-02

TABLE A-16. SAMPLE CALCULATIONS FOR EQUALIZATION BASINS

Component: Equalization basin

(1) Aerated Basin

$A = 9{,}290\ m^2$	$J = 3$ lb O_2/hr-hp
$d_e = 108.7$ m	$O_t{}^\bullet = 0.83$
depth = 2.895 m	Fraction agitated = 0.24
F/D = 37.56	Submerged air flow = 0 m^3/s
$Q = 0.063\ m^3/s$	$N_I = 6$
t = 5 days	d = 61 cm
T = 25°C	w = 126 rad/s
$U_{10} = 4.47$ m/s	

Compound: Toluene

C_L = 10 ppm = 1×10^{-5} g/cm^3

MW = 92.0 g/gmol

$D_w = 8.60 \times 10^{-6}$ cm^2/s

$D_a = 8.70 \times 10^{-2}$ cm^2/s

$H = 6.68 \times 10^{-3}$ atm-m^3/gmol

$D_{ether} = 8.5 \times 10^{-6}$ cm^2/s

Water and Air Properties:

p_w = 1 g/cm^3

$u_w = 8.93 \times 10^{-3}$ g/cm-s

$p_a = 1.20 \times 10^{-3}$ g/cm^3

$u_a = 1.81 \times 10^{-4}$ g/cm-s

$D_{O2,w} = 2.5 \times 10^{-5}$ cm^2/s

TABLE A-16. SAMPLE CALCULATIONS FOR EQUALIZATION BASINS (Continued)

MW_a = 29 g/gmol

MW_w = 18 g/gmol

1. Calculate liquid mass transfer coefficient, k_L:

k_L (m/s) = $[8.22 \times 10^{-9}\ J(POWR)\ (1.024)^{T-20}\ (O_t \times 10^6)\ (MW_w)/(Va_v\rho_w)]$

$(D_w/D_{O2,w})^{0.5}$

a. Calculate POWR:

POWR = 0.5 hp/1000 ft^3 x 35.31 ft^3/m^3 (A x depth)

POWR = 0.0005 hp/ft^3 x 35.31 ft^3/m^3 (9290 m^2 x 2.895 m)

POWR = 475 hp

b. Calculate V:

V = Volume x fraction agitated

V = (100000 ft^2 x 9.5 ft) (0.24)

V = 228000 ft^3

c. Calculate a_v:

a_v = area/volume

a_v = 100000 ft^2/(950000 ft^3)

a_v = 0.1053 ft^{-1}

d. Calculate k_L:

k_L (m/s) = $[8.22 \times 10^{-9}$ (3 lb O_2/hr-hp) (475 hp) $(1.024)^5$

(0.83×10^6) (18 g/gmol)/[(228000 ft^3) (0.1053 ft^{-1})

(1 g/cm^3)]] $[(8.60 \times 10^{-6}\ cm^2/s)/(2.5 \times 10^{-5}\ cm^2/sec)]^{0.5}$

k_L (m/s) = (1.318×10^{-5}) (622) (0.586) = 4.80×10^{-3} m/s

TABLE A-16. SAMPLE CALCULATIONS FOR EQUALIZATION BASINS (Continued)

2. Calculate vapor mass transfer coefficient, k_g:

k_g (m/s) = $1.35 \times 10^{-7} (Re)^{1.42} (P)^{0.4} (Sc_G)^{0.5} (Fr)^{-0.21} (D_a MW_a/d)$

a. Calculate Re (Reynold's Number):

$Re = d^2 wp_a/u_a$

$Re = (61\ cm)^2 (126\ rad/s) (1.20 \times 10^{-3}\ g/cm^3)/1.81 \times 10^{-4}\ g/cm\text{-}s$

$Re = 3.11 \times 10^6$

b. Calculate P (Power Number):

$P = P_I\ g_c/(p_v d^{*5} w^3)$

where

$P_I = 0.85\ (POWR)\ (550\ ft\text{-}lb_f/s\text{-}hp)/N_I$

$P_I = 0.85\ (475\ hp)\ (550\ ft\text{-}lb_f/s\text{-}hp)/6$

$P_I = 37000\ ft\text{-}lb_f/s$

so,

$P = (37000\ ft\text{-}lb_f/s)(32.17\ lb_m\ ft/s^2 - lb_f)/[(1g/cm^3)(28317\ cm^3/ft^3)$

$(lb_m/453.6\ g)\ (61cm(ft/30.48\ cm))^5\ (126\ rad/s)^3]$

$p = (1190290\ lb_m/s^3)/(4.009 \times 10^9\ lb_m/s^3)$

$p = 2.97 \times 10^{-4}$

c. Calculate Sc_G (Schmidt Number):

$Sc_G = u_a/(p_a D_a)$

$Sc_G = (1.81 \times 10^{-4}\ g/cm\text{-}s)/[(1.20 \times 10^{-3}\ g/cm^3)\ (8.70 \times 10^{-2}\ cm^2/s)]$

$Sc_G = 1.734$

TABLE A-16. SAMPLE CALCULATIONS FOR EQUALIZATION BASINS (Continued)

2. (Continued)

d. Calculate Fr (Froude Number):

$Fr = d^* w^2/g_c$

$Fr = (61\ cm/30.48\ cm/ft)\ (126\ rad/s)^2/(32.17\ lb_m\text{-}ft/lb_f\text{-}s^2)$

$Fr = 987.65$

e. Calculate k_g:

$k_g\ (m/s) = 1.35 \times 10^{-7}\ (Re)^{1.42}\ (p)^{0.4}\ (Sc_G)^{0.5}\ (Fr)^{-0.21}\ D_a MW_a/d$

$k_g\ (m/s) = 1.35 \times 10^{-7}\ (3.11 \times 10^6)^{1.42}\ (2.97 \times 10^{-4})^{0.4}\ (1.734)^{0.5}$

$(987.65)^{-0.21}\ (8.70 \times 10^{-2}\ cm^2/s)\ (29\ g/gmol)/(61\ cm)$

$k_g = 1.11 \times 10^{-1}\ m/s$

3. Calculate overall mass transfer coefficient, K:

$1/K = 1/k_L + 1/(K_{eq} k_g)$

a. Calculate K_{eq}:

$K_{eq} = H/RT$

$K_{eq} = (6.68 \times 10^{-3}\ atm\text{-}m^3/gmol)/[(8.21 \times 10^{-5}\ atm\text{-}m^3/gmol\text{-}°K)\ (298°K)]$

$K_{eq} = 0.273$

b. Calculate K:

$1/K = (4.80 \times 10^{-3}\ m/s) + [1/(0.273)\ (1.11 \times 10^{-1}\ m/s)]$

$1/K = 208 + 33$

$K = 4.15 \times 10^{-3}\ m/s$

4. Calculate overall mass transfer coefficient for combined quiescent and turbulent areas, K (weighted by area):

Turbulent area = $0.24 \times A = 0.24\ (9290\ m^2)$

Turbulent area = $2230\ m^2$

$K_T = 4.15 \times 10^{-3}\ m/s$

TABLE A-16. SAMPLE CALCULATIONS FOR EQUALIZATION BASINS (Continued)

4. (Continued)
Quiescent area = (Total Area-Turbulent Area) = 9290 - 2230

Quiescent area = 7060 m^2

K_Q (m/s) = 4.52×10^{-6}

K (m/s) = [(4.52×10^{-6} m/s) (7060 m^2) + (4.15×10^{-3} m/s) (2230 m^2)]/

(2230 m^2 + 7060 m^2)

K = 9.98×10^{-4} m/s

5. Calculate concentration of toluene at vapor-liquid interface, C_L:

$C_L = QC_i/(KA + Q)$

C_L = (0.063 m^3/s) (1×10^{-5} g/cm^3) (10^6 cm^3/m^3)/

[(9.98×10^{-4} m/s) (9290 m^2) + (0.063 m^3/s)]

C_L = 0.0675 g/m^3

6. Calculate air emissions, N_a:

$(N_a)_T$ (Mg/yr) = K C_L A

$(N_a)_T$ (Mg/yr) = (9.98×10^{-4} m/s) (3600 s/hr) (7200 hr/yr) (0.0675 g/m^3)

(9290 m^2) (10^{-6} g/Mg)

$(N_a)_T$ = 16.22 Mg/yr

7. Calculate fraction of toluene emitted from an aerated equalization basin, fe:

$(fe)_T = N_a/QC_i$

$(fe)_T$ = (16.22 Mg/yr)/[(0.063 m^3/s) (3600 s/hr) (7200 hr/yr)

(1×10^{-5} g/cm^3) (10^6 cm^3/m^3) (10^{-6} Mg/g)]

$(fe)_T$ = 0.993

TABLE A-16. SAMPLE CALCULATIONS FOR EQUALIZATION BASINS (Continued)

Component: Equalization basin

(2) Non-Aerated Basin

$A = 9{,}290\ m^2$

$d_e = 108.7\ m$

depth = 2.895

F/D = 37.56

$Q = 0.063\ m^3/s$

t = 5 days

T = 25°C

$U_{10} = 4.47\ m/s$

Compound: Toluene

$C_L = 10\ ppm = 1 \times 10^{-5}\ g/cm^3$

$M_w = 92.0\ g/gmol$

$D_w = 8.60 \times 10^{-6}\ cm^2/s$

$D_a = 8.70 \times 10^{-2}\ cm^2/s$

$H = 6.68 \times 10^{-3}\ atm\text{-}m^3/gmol$

$D_{ether} = 8.5 \times 10^{-6}\ cm^2/s$

Water and Air Properties:

$p_w = 1\ g/cm^3$

$u_w = 8.93 \times 10^{-3}\ g/cm\text{-}s$

$p_a = 1.20 \times 10^{-3}\ g/cm^3$

$u_a = 1.81 \times 10^{-4}\ g/cm\text{-}s$

TABLE A-16. SAMPLE CALCULATIONS FOR EQUALIZATION BASINS (Continued)

1. Calculate liquid mass transfer coefficient, k_L:

k_L (m/s) = $[2.605 \times 10^{-9}$ (F/D) + $1.277 \times 10^{-7}]$ $(U_{10})^2$ $[D_w/D_{ether}]^{0.67}$

k_L (m/s) = $[2.605 \times 10^{-9}$ (37.56) + $1.277 \times 10^{-7}]$ $(4.47\ m/s)^2$

$[(8.6 \times 10^{-6} cm^2/s)/(8.5 \times 10^{-6} cm^2/s)]^{0.67}$

$k_L = 4.53 \times 10^{-6}$ m/s

2. Calculate vapor mass transfer coefficient, k_g:

k_g (m/s) = 4.82×10^{-3} $(U_{10})^{0.78}$ $(Sc_G)^{-0.67}$ $(d_e)^{-0.11}$

a. Calculate Sc_G:

$Sc_G = u_a/(p_a D_a)$

Sc_G = $(1.81 \times 10^{-4}$ g/cm-s)/[$(1.20 \times 10^{-3}$ g/cm^3) $(8.70 \times 10^{-2}$ cm^2/s)]

$Sc_G = 1.734$

b. Calculate k_g:

k_g (m/s) = $4.82 \times 10^{-3} \times (4.47\ m/s)^{0.78}$ $(1.734)^{-0.67}$ $(108.7\ m)^{-0.11}$

k_g (m/s) = $4.82 \times 10^{-3} \times$ (3.215) (0.692) (0.597)

$k_g = 0.0064$ m/s

3. Calculate overall mass transfer coefficient, K:

$1/K = 1/k_L + 1/(K_{eq} k_g)$

a. Calculate K_{eq}:

$K_{eq} = H/RT$

K_{eq} = $(6.68 \times 10^{-3}$ atm-m^3/gmol)/[$(8.21 \times 10^{-5}$ atm-m^3/gmol-°K) (298°K)]

$K_{eq} = 0.273$

b. Calculate K:

$1/K = 1/4.53 \times 10^{-6}$ m/s + [1/(0.273) (0.0064 m/s)]

$1/K = 220750 + 572.3$

$K = 4.52 \times 10^{-6}$ m/s

TABLE A-16. SAMPLE CALCULATIONS FOR EQUALIZATION BASINS (Continued)

4. Calculate concentration of toluene at vapor-liquid interface, C_L:

$$C_L = Q\ C_1/KA + Q$$

$$C_L = [(0.063\ m^3/s)\ (1 \times 10^{-5}\ g/cm^3)\ (10^6\ cm^3/m^3)]/$$
$$[(4.52 \times 10^{-6}\ m/s)\ (9290\ m^2) + (0.063\ m^3/s)]$$

$$C_L = 6.00\ g/m^3$$

5. Calculate air emissions, N_a:

$$(N_a)_T\ (Mg/yr) = K\ C_L\ A$$

$$(N_a)_T\ (Mg/yr) = (4.52 \times 10^{-6}\ m/s)\ (3600\ s/hr)\ (7200\ hr/yr)$$
$$(6.00\ g/m^3)(9290\ m^2)\ (10^{-6}\ Mg/g)$$

$$(N_a)_T = 6.532\ Mg/yr$$

6. Calculate fraction of toluene emitted from a non-aerated equalization basin, fe:

$$(fe)_T = (N_a)_T/QC_1$$

$$(fe)_T = 6.532\ Mg/yr/[(0.063\ m^3/s)\ (3600\ s/hr)\ (7200\ hr/yr)$$
$$(1 \times 10^{-5}\ g/cm^3)\ (10^6\ cm^3/m^3)\ (10^{-6}\ Mg/g)]$$

$$(fe)_T = 0.400$$

TABLE A-17. ESTIMATION TECHNIQUE FOR CLARIFIERS

I. Equations used for calculating liquid and vapor mass transfer coefficients:

<u>Liquid Phase</u>, k_L:

Mackay and Yeun
(for F/D <14 and U_{10} >3.25 m/s):

$$k_L = 1.0 \times 10^{-6} + 144 \times 10^{-4} \times (U^*)^{2.2} \times (Sc_L)^{-0.5},$$

where

$$U^* < 0.3$$

where

$$U^* \text{ (m/s)} = 0.01 \times U_{10} \times (6.1 + 0.63 \times U_{10})^{0.5}$$

and

$$Sc_L = u_L/(p_L D_w)$$

<u>Vapor Phase</u>, k_G:

Mackay and Matasugu:

$$k_G \text{ (m/s)} = 4.82 \times 10^{-3} \times (U_{10})^{0.78} \times (Sc_G)^{-0.67} \times (d_e)^{-0.11}$$

II. Equation for calculating the overall mass transfer coefficient, K:

$$1/K = 1/k_L + 1/(K_{eq} k_G)$$

where

$$K_{eq} = \text{Equilibrium constant} = H/RT$$

III. Equation for calculating the bulk concentration of the organic compound:

$$C_L \text{ (g/m}^3\text{)} = QC_o/(KA + Q)$$

IV. Equation for calculating air emissions, N_a:

$$N_a \text{ (Mg/yr)} = K\ C_L\ A$$

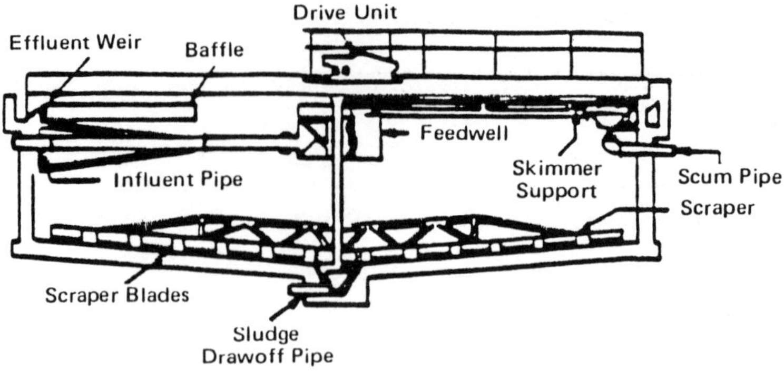

Design Parameter	Range	Typical Design
Diameter (m)	6.1 - 30.5	18.3
Depth (m)	2.4 - 4.5	3.5
Retention time (hr)	1.5 - 7	4.0

Figure A-7. Typical clarifier configuration.

TABLE A-18. EMISSION ESTIMATES FOR CLARIFIERS

Compound	Henry's Law Constant H, atm - m^3/gmol (T = 25°C)	Fraction Emitted, Fe
1,3 Butadiene	1.42E-01	2.8E-02
Toluene	6.68E-03	2.5E-02
Naphthalene	1.18E-03	2.4E-02
1-Butanol	8.90E-06	7.8E-03
Phenol	4.54E-07	5.6E-04

TABLE A-19. SAMPLE CALCULATIONS FOR CLARIFIERS

Component: <u>Clarifier</u>

$A = 262.7\ m^2$

$d_e = 18.29\ m$

depth = 3.505 m

F/D = 5.22

$Q = 0.0639\ m^3/s$

t = 4 hr

T = 25°C

$U_{10} = 4.47\ m/s$

Compound: <u>Toluene</u>

$C_i = 10\ ppm = 1 \times 10^{-5}\ g/cm^3$

MW = 92.0 g/gmol

$D_w = 8.60 \times 10^{-6}\ cm^2/s$

$D_a = 8.70 \times 10^{-2}\ cm^2/s$

$H = 6.68 \times 10^{-3}\ atm\text{-}m^3/gmol$

$D_{ether} = 8.5 \times 10^{-6}\ cm^2/s$

Water and Air Properties:

p_w = [illegible] g/cm^3

$u_w = 8.93 \times 10^{-3}\ g/cm\text{-}s$

$p_a = 1.20 \times 10^{-3}\ g/cm^3$

$u_a = 1.81 \times 10^{-4}\ g/cm\text{-}s$

TABLE A-19. SAMPLE CALCULATIONS FOR CLARIFIERS (Continued)

1. Calculate liquid mass transfer coefficient, k_L:

a. Calculate U*:

U^* (m/s) = $0.01 \times U_{10} (6.1 + 0.63 \times U_{10})^{0.5}$

U^* (m/s) = 0.01×4.47 m/s $(6.1 + 0.63 \times 4.47 \text{ m/s})^{0.5}$

U^* = 0.1335 m/s

b. Calculate Sc_L:

$Sc_L = u_w/(p_w D_w)$

$Sc_L = 8.93 \times 10^{-3}$ g/cm-s/[(1 g/cm^3) (8.60×10^{-6} cm^2/s)]

Sc_L = 1,038

c. Calculate k_L:

k_L (m/s) = $1.0 \times 10^{-6} + 144 \times 10^{-4} (U^*)^{2.2} (Sc_L)^{-0.5}$

k_L (m/s) = $1.0 \times 10^{-6} + 144 \times 10^{-4} (0.1335)^{2.2} \times (1038)^{-0.5}$

$k_L = 6.32 \times 10^{-6}$ m/s

2. Calculate vapor mass transfer coefficient, k_g:

k_g (m/s) = $4.82 \times 10^{-3} \times (U_{10})0.78\ (Sc_G)^{-0.67} (d_e)^{-0.11}$

a. Calculate Sc_G:

$Sc_G = u_a/(p_a D_a)$

Sc_G = (1.81×10^{-4} g/cm-s)/[(1.20×10^{-3} g/cm^3) (8.70×10^{-2} cm^2/s)]

Sc_G = 1.734

b. Calculate k_g:

k_g (m/s) = $4.82 \times 10^{-3} \times (4.47 \text{ m/s})^{0.78} (1.734)^{-0.67} (18.29 \text{ m})^{-0.11}$

k_g (m/s) = $4.82 \times 10^{-3} \times (3.215) (0.692) (0.726)$

k_g = 0.0078 m/s

TABLE A-19. SAMPLE CALCULATIONS FOR CLARIFIERS (Continued)

3. Calculate overall mass transfer coefficient, K:

$1/K = 1/k_L + 1/(K_{eq}k_g)$

a. Calculate K_{eq}:

$K_{eq} = H/RT$

$K_{eq} = 6.68 \times 10^{-3}$ atm-m^3/gmol/[(8.21 x 10^{-5} atm-m^3/gmol-°K) (298°K)]

$K_{eq} = 0.273$

b. Calculate K:

$1/K = 1/k_L + 1/(K_{eq}k_g)$

$1/K$ = (1/6.32 x 10^{-6} m/s) + [1/(0.273) (0.0078 m/s)]

$1/K$ = 158228 + 469.6

K = 6.30 x 10^{-6} m/s

4. Calculate concentration of toluene at vapor-liquid interface, C_L:

$C_L = Q\ C_i/(KA + Q)$

C_L = (0.0639 m^3/s) (1 x 10^{-5} g/cm^3) (106 cm^3/m^3)/

[(6.30 x 10^{-6} m/s) (262.7 m^2) + (0.0639 m^3/s)]

C_L = 9.75 g/m^3

5. Calculate air emissions, N_a:

$(N_a)_T$ (Mg/yr) = K C_L A

$(N_a)_T$ (Mg/yr) = (6.30 x 10^{-6} m/s) (3600 s/hr) (7200 hr/yr)

(9.75 g/m^3) (262.7 m^2) (10^{-6} Mg/g)

$(N_a)_T$ = 0.418 Mg/yr

6. Calculate fraction of toluene emitted from a clarifier, fe:

$(fe)_T = (N_a)_T/QC_i$

$(fe)_T$ = (0.418 Mg/yr)/[(0.0639 m^3/s) (3600 s/hr) (7200 hr/yr)

(1 x 10^{-5} g/cm^3) (10^6cm^3/m^3) (10^{-6} Mg/g)]

(fe)T = 0.0253

Table A-20. For aerated basins, the overall mass transfer coefficient is calculated by considering that the liquid surface is made up of both turbulent and quiescent areas. This overall mass transfer coefficient is estimated using the individual liquid and gas phase coefficients for both areas. Biodegradation is also included in the emissions model as a competing mechanism since biodegradation and volatilization are both significant organic compound removal mechanisms from these units. The removal rate by the competing biodegradation mechanism is more difficult to predict because of its strong dependency on the organic compound characteristics, basin design and operating parameters, and the influent organic compound concentration. For these emissions estimates, the effluent concentration is calculated from the organic compound biorate constant, the active biomass, and the influent concentration. The fraction of the compound emitted is then calculated using this effluent concentration and the overall mass transfer coefficient.

A typical aerated biological treatment basin is shown in Figure A-8.[25] Emission estimates for the aerated basin are based on the typical design dimensions shown in the figure. Table A-21 presents emission rates for typical aerated and non-aerated biological treatment basins. The emission rates are based on influent organic compound concentrations of 40 g/m^3 for each of the five organic compounds. The wastewater is assumed to be at ambient temperature for the purposes of the calculation. Example calculations for both aerated and non-aerated biological treatment basins are shown in Table A-22.

A.1.2.7 Emissions from Treatment Tanks

The technique used to estimate emissions from flocculation and pH adjustment tanks is presented in Table A-23. Individual mass transfer coefficients are estimated based on correlations used for quiescent surface impoundments. Overall mass transfer coefficients are estimated based on values obtained for the individual coefficients. The wastewater flowing through the tank is assumed to be well-mixed. Therefore, the effluent concentration is the driving force for air emissions. The overall coefficients, the liquid surface area in tanks, and the wastewater effluent concentrations for each organic compound are then multiplied together to estimate the emission rate of each organic compound.

TABLE A-20. ESTIMATION TECHNIQUE FOR BIOLOGICAL TREATMENT BASINS

(1) Aerated Basin

I. Equations used for calculating liquid and vapor mass transfer coefficients:

Liquid Phase, k_L:

Thibodeaux:

$$k_L \text{ (m/s)} = [8.22 \times 10^{-9} J \text{ (POWR)}(1.024)^{T-20} O_t \, 10^6 MW_L/(Va_v \rho_w)] \, (D_w/D_{O2,w})^{0.5}$$

Gas Phase

Reinhardt:

$$k_G \text{ (m/s)} = 1.35 \times 10^{-7} (Re)^{1.42} (P)^{0.4} (Sc_G)^{0.5} (Fr)^{-0.21} (D_a MW_a/d)$$

II. Equation for calculating the overall mass transfer coefficient, K;

$$1/K = 1/(k_L) + 1/(K_{eq} k_G)$$

where

K_{eq} = Equilibrium constant = H/RT

Equation for calculating overall mass transfer coefficient for combined quiescent and turbulent areas:

$$K = K_Q A_Q + K_T A_T)/(A_Q + A_T)$$

where

K_Q = overall mass transfer coefficient for quiescent area,

K_T = overall mass transfer coefficient for turbulent area,

A_Q = quiescent surface area, and

A_T = turbulent surface area.

TABLE A-20. ESTIMATION TECHNIQUE FOR BIOLOGICAL TREATMENT BASINS (Continued)

III. Equation for calculating the effluent concentration for a well-mixed system:

$$a = K' = (KA/Q) + 1$$

$$b = K_s K'' + (V/Q)\, K_{max} b_i - C_o$$

$$c = -K_s C_o$$

$$C_L = [-b + (b^2 - 4ac)^{0.5}]/(2a)$$

where

K_{max} = maximum biorate, g/s-g biomass

K_s = half saturation constant, g/m^3

b_i = Active biomass, g/m^3

C_o = inlet organic compound concentration, g/m^3

C_L = effluent organic compound concentration, g/m^3

IV. Equation for calculating the fraction emitted from a well-mixed system:

$$f_e = KAC_L/QC_o$$

(2) Non-Aerated Basins

I. Equations used for calculating liquid and vapor mass transfer coefficients:

Liquid Phase, k_L:

Springer et al (for 14 <F/D <51.2 and U_{10} >3.25 m/s):

$$k_L \text{ (m/s)} = [2.605 \times 10^{-9} \text{ (F/D)} + 1.277 \times 10^{-7}] \times (U_{10})^2 \times (D_w/D_{ether})^{2/3}$$

Vapor Phase, k_G:

MacKay and Matasugu:

$$k_G \text{ (m/s)} = 4.82 \times 10^{-3} \times (U_{10})^{0.78} \times (Sc_G)^{-0.67} \times (d_e)^{-0.11}$$

TABLE A-20. ESTIMATION TECHNIQUE FOR BIOLOGICAL TREATMENT BASINS (Continued)

II. Equation for calculating the overall mass transfer coefficient, K:

$$1/K = 1/k_L + 1/(K_{eq}k_G)$$

where

K_{eq} = Equilibrium constant = H/RT

III. Equation for calculating the effluent concentration for a well-mixed system:

$$a = K' = (KA/Q) + 1$$

$$b = K_sK' + (V/Q)\ K_{max}b_i - C_o$$

$$c = -K_sC_o$$

$$C_L = [-b + (b^2 - 4ac)^{0.5}]/(2a)$$

where

K_{max} = maximum biorate, g/s-g biomass

K_s = half saturation constant, g/m^3

b_i = Active biomass, g/m^3

C_o = inlet organic compound concentration, g/m^3

C_L = effluent organic compound concentration, g/m^3

IV. Equation for calculating the fraction emitted from a well-mixed system:

$$f_e = KAC_L/QC_o$$

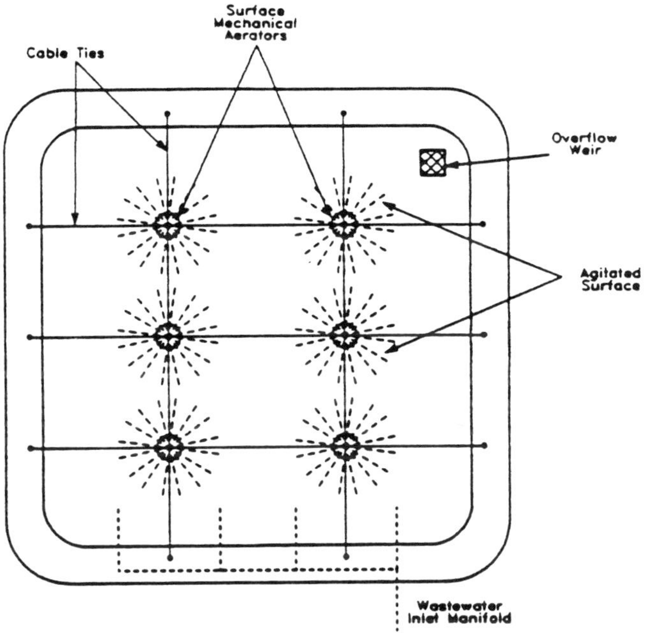

Design Parameter	Typical Design
Effective Diameter (m)	150
Surface Area (m^2)	17,650
Water depth (m)	2.0
Retention time (days)	6.5

Figure A-8. Typical Aerated Biological Treatment Basin

TABLE A-21. EMISSION ESTIMATES FOR BIOLOGICAL TREATMENT BASINS

Compound	Fraction Emitted, Fe (T=25°C)
Aerated	
1,3-Butadiene	2.9E-01
Toluene	1.2E-01
Naphthalene	1.7E-01
1-Butanol	4.1E-02
Phenol	1.9E-05
Non-Aerated	
1,3-Butadiene	2.2E-01
Toluene	6.0E-02
Naphthalene	1.2E-01
1-Butanol	3.0E-01
Phenol	3.9E-04

TABLE A-22. SAMPLE CALCULATIONS FOR BIOLOGICAL TREATMENT BASINS

I. Component: Aerated Basin Containing Biomass

$A = 17{,}652\ m^3$	$J = 3\ lb\ O_2/hr\text{-}hp$
$de = 149.9\ m$	$O_t = 0.83$
$depth = 1.981\ m$	$N_I = 8$
$V = 34{,}972\ m^3$	$d = 61\ cm$
$F/D = 75.67$	$w = 126\ rad/s$
$Q = 0.0623\ m^3/s$	Fraction agitated = 0.24
$t = 6.5$ days	Submerged air flow = $0\ m^3/s$
$T = 25°C$	bi = Active biomass = 4 g/l
$U_{10} = 4.47\ m/s$	Biomass solids = $0.00\ m^3/s$

Compound: Toluene

$C_o = 40\ ppm = 40\ g/m^3$

$MW = 92.0\ g/gmol$

$Dw = 8.60 \times 10^{-6}\ cm^3/s$

$M_v = 8.70 \times 10^{-2}\ cm^3/s$

$H = 6.68 \times 10^{-3}\ atm\text{-}m^3/gmol$

$D_{ether} = 8.50 \times 10^{-6}\ cm^3/s$

$K_{max} = 73.5\ mg/gbiomass\text{-}hr = 2.04 \times 10^{-5}\ g/gbiomass/s$

$K_s = 30.6\ mg/l = 30.6\ g/m^3$

Water and Air Properties:

$p_w = 1\ g/cm^3$

$u_w = 8.93 \times 10^{-3}\ g/cm\text{-}s$

$u_a = 1.81 \times 10^{-4}\ g/cm\text{-}s$

$D_{O2,w} = 2.5 \times 10^{-5}\ cm^3/s$

$MW_a = 29\ g/gmol$

TABLE A-22. SAMPLE CALCULATIONS FOR BIOLOGICAL TREATMENT BASINS (Continued)

1. Calculate liquid mass transfer coefficient, k_L:

$$k_L\ (m/s) = [8.22 \times 10^{-9}\ J(POWR)(1.024)^{T-20}(O_t \times 10^6)\ (MW_w/(Va_vP_w))]\ (D_w/D_{O2,w})^{0.5}$$

a. Calculate POWR:

$$POWR\ (hp) = 0.5\ hp/1000\ ft^3 \times 35.31\ ft^3/m^3\ (A \times depth)$$

$$POWR\ (hp) = 0.0005\ hp/ft^3 \times 35.31\ ft^3/m^3\ (17652\ m^3 \times 1.9812\ m)$$

$$POWR = 617.43\ hp$$

b. Calculate V:

$$V\ (ft^3) = Volume \times fraction\ agitated$$

$$V\ (ft^3) = 34{,}972\ m^3\ (3.28\ ft^3/m^3)$$

$$V = 296180\ ft^3$$

c. Calculate a_v:

$$a_v = area/volume,\ ft^{-1}$$

$$a_v = 17{,}652\ m^3/[(34{,}972\ m^3) \times (3.28\ ft/m)]$$

$$a_v = 0.1538\ ft^{-1}$$

d. Calculate k_L:

$$k_L\ (m/s) = [8.22 \times 10^{-9}\ (3\ lb\ O_2/hr\text{-}hp)\ (617.43\ hp)\ (1.024)^5]$$
$$[0.83 \times 10^6 \times 18\ g/gmol/[(296180\ ft^3)\ (0.1538\ ft^{-1})\ (1\ g/cm^3)]]$$
$$[8.60 \times 10^{-6}\ cm^3/s)/(2.50 \times 10^{-5}\ cm^3/s)]^{0.5}$$

$$k_L\ (m/s) = (1.71 \times 10^{-5})\ (327.97)\ (0.5865)$$

$$k_L = 0.00329\ m/s$$

TABLE A-22. SAMPLE CALCULATIONS FOR BIOLOGICAL TREATMENT BASINS (Continued)

2. Calculate vapor mass transfer coefficient, k_g:

k_g (m/s) = $1.35 \times 10^{-7} (Re)^{1.42} (P)^{0.4} (Sc_G)^{0.5} (Fr)^{-0.21} (D_a MW_a/d)$

a. Calculate Re (Reynold's Number):

$Re = d^2 w p_a/u_a$

$Re = (61\ cm)^2\ (126\ rad/s)\ (1.20 \times 10^{-3}\ g/cm^3)/(1.81 \times 10^{-4}\ g/cm\text{-}s)$

$Re = 3.11 \times 10^6$

b. Calculate P (Power Number):

$P = P_I\ g_c/(p_w d^{*5} w^3)$

where

$P_I = 0.85\ (POWR)\ (550\ ft\text{-}lb_f/s\text{-}hp)/N_I$

$P_I = 0.85\ (475\ hp)\ (550\ ft\text{-}lb_f/s\text{-}hp)/8$

$P_I = 36{,}081\ ft\text{-}lb_f/s$

so,

$P = (36{,}081\ ft\text{-}lb_f/s)(32.17\ lb_m\ ft/s^2 - lb_f)/$

$[(1 g/cm^3)\ (lb_m/453.6\ g)\ (28317\ cm^3/ft^3)$

$[(61 cm)(ft/30.48\ cm)]^5\ (126\ rad/s)^3]$

$P = 2.895 \times 10^{-4}$

c. Calculate Sc_G (Schmidt Number):

$Sc_G = u_a/(p_a D_a)$

$Sc_G = (1.81 \times 10^{-4}\ g/cm\text{-}s)/[(1.20 \times 10^{-3}\ g/cm^3)\ (8.70 \times 10^{-2}\ cm^3/s)]$

$Sc_G = 1.734$

d. Calculate Fr (Froude Number):

$Fr = d^*\ w^2/g_c$

$Fr = (61\ cm/30.48\ cm/ft)\ (126\ rad/s)^2/(32.17\ lb_m\text{-}ft/lb_f\text{-}s^2)$

$Fr = 987.65$

TABLE A-22. SAMPLE CALCULATIONS FOR BIOLOGICAL TREATMENT BASINS (Continued)

2. (Continued)

e. Calculate k_g:

$$k_g \text{ (m/s)} = 1.35 \times 10^{-7} (3.11 \times 10^{6})^{1.42} (2.895 \times 10^{-4})^{0.4} (1.734)^{0.5} (987.65)^{-0.21} (8.70 \times 10^{-2} \text{ cm}^3\text{/s}) (29 \text{ g/gmol})/(61 \text{ cm})$$

$$k_g = 0.110 \text{ m/s}$$

3. Calculate overall mass transfer coefficient, K:

$$1/K = 1/k_L + 1/(K_{eq}k_g)$$

a. Calculate K:

$$K = H/RT$$

$$K = (6.68 \times 10^{-3} \text{ atm-m}^3\text{/gmol})/[(8.21 \times 10^{-5} \text{ atm-m}^3\text{/gmol-°K}) (298°\text{K})]$$

$$K = 0.273$$

b. Calculate K:

$$1/K = 1/k_L + 1/(K_{eq}k_g)$$

$$1/K = (1/0.00329 \text{ m/s}) + [1/(0.273) (0.110 \text{ m/s})]$$

$$1/K = 303.95 + 33.30$$

$$K = 0.00296$$

4. Calculate overall mass transfer coefficient for combined quiescent and turbulent areas, K (weighted by area):

Turbulent area = 4236.5 m^3

K = 0.00296

Quiescent area = 13415.5 m^3

a. Calculate quiescent liquid mass transfer coefficient, k_L:

$$k_L \text{ (m/s)} = [2.611 \times 10^{-7} (U_{10})^2 [D_w/D_{ether}]^{2/3}$$

$$k_L \text{ (m/s)} = [2.611 \times 10^{-7} (4.47 \text{ m/s})^2 [(8.60 \times 10^{-6}\text{cm}^2\text{/s})/(8.50 \times 10^{-6}\text{cm}^3\text{/s})]^{0.667}$$

$$k_L = 5.25 \times 10^{-6} \text{ m/s}$$

TABLE A-22. SAMPLE CALCULATIONS FOR BIOLOGICAL TREATMENT BASINS (Continued)

4. (Continued)

b. Calculate quiescent vapor mass transfer coefficient, k_g:

$$k_g\ (m/s) = 4.82 \times 10^{-3}\ (U_{10})^{0.78}\ (Sc_G)^{-0.67}\ (d_e)^{-0.11}$$

$$k_g\ (m/s) = 4.82 \times 10^{-3} \times (4.47\ m/s)^{0.78}\ (1.734)^{-0.67}\ (149.9\ m)^{-0.11}$$

$$k_g\ (m/s) = 4.82 \times 10^{-3} \times (3.215)\ (0.6916)\ (0.5763)$$

$$k_g = 0.00618\ m/s$$

c. Calculate quiescent overall mass transfer coefficient, K:

$$1/K = 1/k_L + 1/(K_{eq}k_g)$$

$$1/K = 1/5.25 \times 10^{-6}\ m/s + [1/(0.273)\ (0.00618\ m/s)]$$

$$1/K = 190480 + 593$$

$$K = 5.24 \times 10^{-6}$$

d. Calculate K:

$$K\ (m/s) = [K_QA_Q + K_TA_T]/[A_Q + A_T]$$

$$K\ (m/s) = [(5.24 \times 10^{-6}\ m/s)\ (13415.5\ m^3) + 2.96 \times 10^{-3}\ m/s)\ (4236.5\ m^3)]/(13415.5\ m^3 + 4236.5\ m^3)$$

$$K = 7.15 \times 10^{-4}\ m/s$$

5. Calculate the effluent concentration for toluene for a well-mixed system, C_L:

$$C_L\ (g/m^3) = [-b + (b^2 - 4ac)^{0.5}]/(2a)$$

a. Calculate a:

$$a = K' = (KA/Q) + 1$$

$$= [(7.15 \times 10^{-4}\ m/s)(17{,}652\ m^2)/(0.0623\ m^3/s)] + 1$$

$$= 203.7$$

TABLE A-22. SAMPLE CALCULATIONS FOR BIOLOGICAL TREATMENT BASINS (Continued)

5. (Continued)

b. Calculate b:

$b = K_s K' + (V/Q) K_{max} bi - C_o$

$= [(30.6 g/m^3)(203.7) + (34,972 m^3/0.0623 m^3/s)$

$(2.04 \times 10^{-5} g/gbiomass\text{-}s)(4.0 g/l)(1,000 l/m^3) - 40 g/m^3]$

$= 6,233 + 45,806 - 40 = 52,000 g/m^3$

c. Calculate c:

$c = -K_s C_o$

$= (-30.6 g/m^3)(40 g/m^3) = -1,224 g^2/m^6$

d. Calculate C_L:

$C_L (g/m^3) = [-52,000 g/m^3 + [(52,000 g/m^3)^2 - 4(203.7)$

$(-1,224 g^2/m^6)]^{0.5}]/(2(203.7))$

$= 0.02352 g/m^3$

e. Calculate the fraction emitted for a well-mixed system, fe:

$fe = KAC_L/QC_o$

$= (7.15 \times 10^{-4} m/s)(17,652 m^3)(0.02352 g/m^3)/[(0.0623 m^3/s)(40 g/m^3)]$

$= 0.119$

II. Component: <u>Non-Aerated Basin Containing Biomass</u>

$A = 1,500 m^2$ bi = Active biomass = 0.05 g/l

$d_e = 43.7 m$ Biomass solids = 0.00 m^3/s

depth = 1.8 m

$V = 2,700 m^3$

F/D = 24.28

TABLE A-22. SAMPLE CALCULATIONS FOR BIOLOGICAL TREATMENT BASINS (Continued)

$Q = 0.00156\ m^3/s$

$t = 20.03$ days

$T = 25°C$

$U_{10} = 4.47\ m/s$

Compound: Toluene

$C_o = 40$ ppm $= 40\ g/m^3$

$MW = 92.0$ g/gmol

$Dw = 8.60 \times 10^{-6}\ cm^3/s$

$M_v = 8.70 \times 10^{-2}\ cm^3/s$

$H = 6.68 \times 10^{-3}$ atm-m^3/gmol

$D_{ether} = 8.5 \times 10^{-6}\ cm^3/s$

$K_{max} = 73.5$ mg/gbiomass-hr $= 2.04 \times 10^{-5}$ g/gbiomass/s

$K_s = 30.6$ mg/l $= 30.6\ g/m^3$

Water and Air Properties:

$p_w = 1\ g/cm^3$

$u_w = 8.93 \times 10^{-3}$ g/cm-s

$p_a = 1.20 \times 10^{-3}\ g/cm^3$

$u_a = 1.81 \times 10^{-4}$ g/cm-s

1. Calculate liquid mass transfer coefficient, k_L:

$$k_L\ (m/s) = [2.605 \times 10^{-9}\ (F/D) + 1.277 \times 10^{-7}]\ (U_{10})^2\ [D_w/D_{ether}]^{0.67}$$

$$k_L\ (m/s) = [2.605 \times 10^{-9}\ (24.28) + 1.277 \times 10^{-7}]\ (4.47\ m/s)^2\ [(8.6 \times 10^{-6} cm^2/s)/(8.5 \times 10^{-6} cm^2/s)]^{0.67}$$

$$k_L = 3.84 \times 10^{-6}\ m/s$$

TABLE A-22. SAMPLE CALCULATIONS FOR BIOLOGICAL TREATMENT BASINS (Continued)

2. Calculate vapor mass transfer coefficient, k_g:

k_g (m/s) = $4.82 \times 10^{-3} (U_{10})^{0.78} (Sc_G)^{-0.67} (de)^{-0.11}$

a. Calculate Sc_G:

$Sc_G = u_a/(p_a D_a)$

$Sc_G = (1.81 \times 10^{-4}$ g/cm-s)/[(1.20×10^{-3} g/cm^3) (8.70×10^{-2} cm^3/s)]

$Sc_G = 1.734$

b. Calculate k_g:

k_g (m/s) = $4.82 \times 10^{-3} \times (4.47$ m/s$)^{0.78} (1.734)^{-0.67} (43.7$ m$)^{-0.11}$

k_g (m/s) = $4.82 \times 10^{-3} \times (3.215) (0.692) (0.660)$

$k_g = 7.08 \times 10^{-3}$ m/s

3. Calculate overall mass transfer coefficient, K:

$1/K = 1/k_L + 1/(K_{eq} k_g)$

a. Calculate K_{eq}:

$K_{eq} = H/RT$

K_{eq} = (6.68×10^{-3} atm-m^3/gmol)/[(8.21×10^{-5} atm-m^3/gmol-°K) (298°K)]

$K_{eq} = 0.273$

b. Calculate K:

$1/K = 1/3.84 \times 10^{-6}$ m/s + [1/(0.273) (7.08×10^{-3} m/s)]

$1/K = 260400 + 517.4$

$K = 3.83 \times 10^{-6}$ m/s

4. Calculate the effluent concentration for toluene for a well-mixed system, C_L:

C_L (g/m^3) = $[-b + (b^2 - 4ac)^{0.5}]/(2a)$

TABLE A-22. SAMPLE CALCULATIONS FOR BIOLOGICAL TREATMENT BASINS (Continued)

4. (Continued)

a. Calculate a:

$$a = K' = (KA/Q) + 1$$

$$= [(3.83 \times 10^{-6}\ m/s)(1{,}500\ m^3)/(0.00156\ m^3/s)] + 1$$

$$= 4.683$$

b. Calculate b:

$$b = K_s\ K' + (V/Q)\ K_{max}\ bi - C_o$$

$$= [(30.6\ g/m^3)(4.683) + (2{,}700\ m^3/0.00156\ m^3/s)$$

$$(2.04 \times 10^{-5}\ g/gbiomass\text{-}s)(0.05\ g/l)(1{,}000\ l/m^3) - 40\ g/m^3]$$

$$= 143.30 + 1{,}765.4 - 40 = 1{,}869\ g/m^3$$

c. Calculate c:

$$c = -K_sC_o$$

$$= (-30.6\ g/m^3)(40\ g/m^3) = -1{,}224\ g^2/m^6$$

d. Calculate C_L:

$$C_L\ (g/m^3) = [-1{,}869\ g/m^3 + [(1{,}869\ g/m^3)^2 - 4(4.683)$$

$$(-1{,}224\ g^2/m^6)]^{0.5}]/2(4.683)$$

$$= 0.6538\ g/m^3$$

e. Calculate the fraction emitted for a well-mixed system, fe:

$$fe = KAC_L/QC_o$$

$$= (3.83 \times 10^{-6}\ m/s)(1{,}500\ m^2)(0.6538\ g/m^3)/[(0.00156\ m^3/s)(40\ g/m^3)]$$

$$= 0.0602$$

TABLE A-23. ESTIMATION TECHNIQUE FOR TREATMENT TANKS

I. Equations used for calculating liquid and vapor mass transfer coefficients:

Liquid Phase, k_L:

Mackay and Yeun
(for F/D <14 and U_{10} >3.25 m/s):

$$k_L = 1.0 \times 10^{-6} + 144 \times 10^{-4} \times (U^*)^{2.2} \times (Sc_L)^{-0.5},$$

where

$$U^* < 0.3$$

where

$$U^* \text{ (m/s)} = 0.01 \times U_{10} \times (6.1 + 0.63 \times U_{10})^{0.5}$$

and

$$Sc_L = u_L/(p_L D_w)$$

Vapor Phase, k_G:

Mackay and Matasugu:

$$k_G \text{ (m/s)} = 4.82 \times 10^{-3} \times (U_{10})^{0.78} \times (Sc_G)^{-0.67} \times (d_e)^{-0.11}$$

II. Equation for calculating the overall mass transfer coefficient, K:

$$1/K = 1/k_L + 1/(K_{eq} k_G)$$

where

$$K_{eq} = \text{Equilibrium constant} = H/RT$$

III. Equation for calculating the bulk concentration of the organic compound:

$$C_L \text{ (g/m}^3\text{)} = QC_o/(KA + Q)$$

IV. Equation for calculating air emissions, N_a:

$$N_a \text{ (Mg/yr)} = K\ C_L\ A$$

Emission estimates based on typical design dimensions for the tanks are shown in Table A-24. Table A-24 presents the fraction emitted for five example organic compounds. The emission rates are based on a wastewater flow rate of 1,000 gpm through the tanks. Each of the five organic compounds are assumed to be present at a concentration of 10 ppm. The wastewater is assumed to be at ambient temperature for the purposes of the calculation. An example calculation for the tanks is shown in Table A-25.

A.1.3 Oil-Water Separators

A theoretical model for predicting VOC emissions from separators has been developed by the Shell Oil Company.[26] The general procedure employed by this model to estimate emissions is presented in Table A-26. The model takes into account wind flowing over the oil/water surface with a logarithmic wind profile. A mass transfer coefficient can be calculated using regressed curves that Shell developed. This mass transfer coefficient is used to calculate the volatilization of each compound from the oil surface into the atmosphere. The model assumes that the volatilization of each component is not affected by the compound matrix in the oil layer. The model also assumes that mass transfer through the liquid is much faster than mass transfer into the gas phase (above the liquid surface). That is, the gas phase resistance is controlling.

A typical oil-water separator is shown in Figure A-9.[27] Emission estimates for the oil-water separator are based on the typical design dimensions shown in the figure. Table A-27 presents the fraction emitted for five example organic compounds. The emission rates are based on a wastewater flow rate of 4,000 gpm through the oil-water separator. Each of the five organic compounds are assumed to be present at a concentration of 500 ppm. The wastewater is assumed to be at 30°C for the purposes of the calculation. An example calculation for oil-water separators is shown in Table A-28.

A.1.4 Weirs

Mass transfer correlations developed from volatilization-reaeration theory were used to estimate emissions from weirs.[25] The general procedure

TABLE A-24. EMISSION ESTIMATES FOR TREATMENT TANKS

Compound	Henry's Law Constant H, atm - m^3/gmol (25°C)	Fraction Emitted, Fe
1,3 Butadiene	1.42E-01	1.0E-02
Toluene	6.68E-03	9.2E-03
Naphthalene	1.18E-03	8.6E-03
1-Butanol	8.90E-06	2.9E-03
Phenol	4.54E-07	2.1E-04

TABLE A-25. SAMPLE CALCULATIONS FOR TREATMENT TANKS

Component: Treatment Tank, (Non-Aerated)

$A = 92.9\ m^2$

$de = 10.9\ m$

$depth = 4.8768\ m$

$F/D = 2.23$

$Q = 0.063\ m^3/s$

$t = 2\ hr$

$T = 25°C$

$U_{10} = 4.47\ m/s$

Compound: Toluene

$C_L = 10\ ppm = 1 \times 10^{-5}\ g/cm^3$

$MW = 92.0\ g/gmol$

$Dw = 8.60 \times 10^{-6}\ cm^2/s$

$D_a = 8.70 \times 10^{-2}\ cm^2/s$

$H = 6.68 \times 10^{-3}\ atm\text{-}m^3/gmol$

$D_{ether} = 8.5 \times 10^{-6}\ cm^2/s$

Water and Air Properties:

$p_w = 1\ g/cm^3$

$u_w = 8.93 \times 10^{-3}\ g/cm\text{-}s$

$p_a = 1.20 \times 10^{-3}\ g/cm^3$

$u_a = 1.81 \times 10^{-4}\ g/cm\text{-}s$

TABLE A-25. SAMPLE CALCULATIONS FOR TREATMENT TANKS (Continued)

1. Calculate liquid mass transfer coefficient, k_L:

a. Calculate U*:

$U^* \text{ (m/s)} = 0.01 \times U_{10} \ (6.1 + 0.63 \times U_{10})^{0.5}$

$U^* \text{ (m/s)} = 0.01 \times 4.47 \text{ m/s} \ (6.1 + 0.63 \times 4.47 \text{ m/s})^{0.5}$

$U^* = 0.1335$ m/s

b. Calculate Sc_L:

$Sc_L = u_w/(p_w D_w)$

$Sc_L = 8.93 \times 10^{-3}$ g/cm-s/[(1 g/cm^3) (8.60 x 10^{-6} cm^2/s)]

$Sc_L = 1{,}038$

c. Calculate k_L:

$k_L \text{ (m/s)} = 1.0 \times 10^{-6} + 144 \times 10^{-4} \ (U^*)^{2.2} \ (Sc_L)^{-0.5}$

$k_L \text{ (m/s)} = 1.0 \times 10^{-6} + 144 \times 10^{-4} \ (0.1335)^{2.2} \times (1038)^{-0.5}$

$k_L = 6.32 \times 10^{-6}$ m/s

2. Calculate vapor mass transfer coefficient, k_g:

$k_g \text{ (m/s)} = 4.82 \times 10^{-3} \times (U_{10})^{0.78} \ (Sc_G)^{-0.67} \ (de)^{-0.11}$

a. Calculate Sc_G:

$Sc_G = u_a/(p_a D_a)$

Sc_G = (1.81 x 10^{-4} g/cm-s)/[(1.20 x 10^{-3} g/cm^3) (8.70 x 10^{-2} cm^2/s)]

$Sc_G = 1.734$

b. Calculate k_g:

$k_g \text{ (m/s)} = 4.82 \times 10^{-3} \times (4.47 \text{ m/s})^{0.78} \ (1.734)^{-0.67} \ (10.9 \text{ m})^{-0.11}$

$k_g \text{ (m/s)} = 4.82 \times 10^{-3} \times (3.215) \ (0.692) \ (0.769)$

$k_g = 0.00824$ m/s

TABLE A-25. SAMPLE CALCULATIONS FOR TREATMENT TANKS
(Continued)

3. Calculate overall mass transfer coefficient, K:

$1/K = 1/k_L + 1/(K_{eq}k_g)$

a. Calculate K_{eq}:

$K_{eq} = H/RT$

$K_{eq} = 6.68 \times 10^{-3}$ atm-m^3/gmol/[(8.21 x 10^{-5} atm-m^3/gmol•K) (298•K)]

$K_{eq} = 0.273$

b. Calculate K:

$1/K = 1/k_L + 1/(K_{eq}k_g)$

$1/K = (1/6.32 \times 10^{-6}$ m/s) + [1/(0.273) (0.00824 m/s)]

$1/K = 158230 + 445$

$K = 6.30 \times 10^{-6}$ m/s

4. Calculate concentration of toluene at vapor-liquid interface, C_L:

$C_L = Q\ C_i/(KA + Q)$

C_L = (0.063 m^3/s) (1 x 10^{-5} g/cm^3) (10^6 cm^3/m^3)/

[(6.30 x 10^{-6} m/s) (92.9 m^2) + (0.063 m^3/s)]

C_L = 9.91 g/m^3

5. Calculate air emissions, Na:

$(N_a)_T$ (Mg/yr) = K C_L A

$(N_a)_T$ (Mg/yr) = (6.30 x 10^{-6} m/s) (3600 s/hr) (7200 hr/yr)

(9.91 g/m^3) (92.9 m^2) (10^{-6} Mg/g)

$(N_a)_T$ = 0.150 Mg/yr

6. Calculate fraction of toluene emitted from a treatment tank, fe:

$(fe)_T = (N_a)_T/QC_i$

$(fe)_T$ = (0.150 Mg/yr)/[(0.063 m^3/s) (3600 s/hr) (7200 hr/yr)

(1 x 10^{-5} g/cm^3) (10^6cm^3/m^3) (10^{-6} Mg/g)]

$(fe)_T$ = 0.00921

TABLE A-26. ESTIMATION TECHNIQUE FOR OIL-WATER SEPARATORS

I. Calculate concentration of the organic compound in bulk liquid, C_L

$$C_L(\text{gmol/m}^3) = QC_i/(MW_i \times Q),$$

II. Calculate partial pressure at oil-air interface, p*

$$p^* = H \times C_L$$

III. Calculate Schmidt number on vapor side Sc_G:

$$Sc_G = u_a/(p_a D_a)$$

IV. Calculate non-dimensional downwind distance, Sqig:

$$Sqig = (0.064 \times U_{10} \times L)\ (D_a)_i$$

V. Calculate curves defined in shell model, k_{ma1} and k_{ma5}

a. $k_{ma1} = 0.328\ Sqig^{-0.227} + 0.298\ Sqig^{-0.127}$, for $Sc_G = 1$

b. $k_{ma5} = 0.431\ Sqig^{-0.129}$ for $Sc_G = 5$

VI. Interpolate between the two curves using calculated Schmidt number, k_m/alpha:

$$k_m/\text{alpha} = k_{ma5} - [(5 - Sc_G)/(5 - 1) \times (k_{ma5} - k_{ma1})]$$

VII. Calculate local mass transfer coefficient, k_m:

$$k_m = (k_m/\text{alpha}) \times \text{alpha}$$

VIII. Calculate concentration of the organic compound at vapor-liquid interface, C_L:

$$C_L\ (\text{gmol/m}^3) = p^*/RT$$

IX. Calculate emission of the organic compound from oil-water separator, $(N_a)i$

$$(N_a)_i = k_m \times C_L \times L \times W,\ \text{mol/s}$$

X. Calculate fraction emitted of the organic compound from oil-water separator, fe

$$(fe) = (N_a)_i/QC_i$$

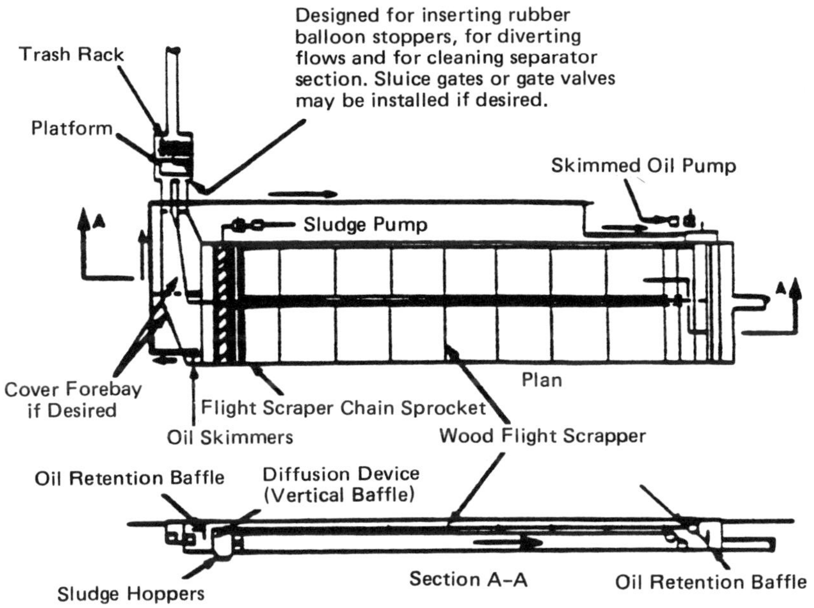

Design Parameter	Range	Typical Design
Separator length (m)	6.1 - 18.0	13.7
Separator width (m)	4.6 - 10.7	7.6
Retention time (hr)	0.6 - 1.0	0.8

Figure A-9. Typical oil/water separator configuration.

TABLE A-27. EMISSION ESTIMATES FOR OIL/WATER SEPARATORS

Compound	Henry's Law Constant H, atm - m^3/gmol (25°C)	Henry's Law Constant H, atm - m^3/gmol (30°C)	Fraction Emitted, Fe (T=30°C)
1,3-Butadiene	1.42E-01	1.65E-01	1.0E+00
Toluene	6.68E-03	8.61E-03	4.7E-01
Naphthalene	1.18E-03	1.73E-03	9.0E-02
1-Butanol	8.90E-06	1.29E-05	6.9E-04
Phenol	4.54E-07	6.90E-07	3.7E-05

TABLE A-28. SAMPLE CALCULATIONS FOR OIL/WATER SEPARATORS

Component: Oil-Water Separator

$L = 1{,}371$ cm

$w = 762$ cm

depth = 244 cm

$Q = 4000$ gal/min

$F_i = QC_i = 126.18$ g/s

$T = 30°C$

$P = 1$ atm

$R = 82.05$ atm-cm^3/gmol

$k_v = 0.0056$ cm^2/s

alpha = 4.3147 cm^2/s

$U_{10} = 447$ cm/s

Compound: Toluene

$C_i = 500$ ppm $= 5 \times 10^{-4}$ g/cm^3 $\quad$ $p_a = 1.20 \times 10^{-3}$ g/cm^3

MW = 92.0 g/gmol $\quad$ $u_a = 1.81 \times 10^{-4}$ g/cm/s

$D_w = 9.74 \times 10^{-6}$ cm^2/s

$D_a = 8.70 \times 10^{-2}$ cm^2/s

$H = 8.61 \times 10^{-3}$ atm-m^3/gmol

1. Calculate concentration of toluene in bulk liquid, C_L:

C_L (gmol/m^3) = F_i (264.17 gal/m^3) (60 s/min)/(MW x Q)

C_L (gmol/m^3) = 126.18 g/s (264.17 gal/m^3) (60 s/min)/
(92.0 g/gmol) (4,000 gal/min)

$C_L = 5.435$ gmol/m^3

TABLE A-28. SAMPLE CALCULATIONS FOR OIL/WATER SEPARATORS (Continued)

2. Calculate partial pressure at interface, p*:

$p^* = H \times C_L$

$p^* = (8.61 \times 10^{-3}\ \text{atm-m}^3/\text{gmol})(5.435\ \text{gmol/m}^3)$

$p^* = 0.0468\ \text{atm}$

3. Calculate Schmidt number on vapor side Sc_G:

$Sc_G = u_a/(p_a D_a)$

$Sc_G = (1.81 \times 10^{-4}\ \text{g/cm/s}) / [(1.20 \times 10^{-3}\ \text{g/cm}^3 \times 0.087\ \text{cm}^2/\text{s})]$

$Sc_G = 1.734$

4. Calculate non-dimensional downward distance, Sqig:

$Sqig = 0.064 \times U_{10} \times L/D_a$

$Sqig = 0.064\ (447\ \text{cm/s})\ (1371\ \text{cm})/(0.087\ \text{cm}^2/\text{s})$

$Sqig = 450{,}823$

5. Calculate curves defined in shell model, k_{ma1} and k_{ma5}:

a. $k_{ma1} = 0.328\ Sqig^{-0.227} + 0.298\ Sqig^{-0.127}$, for $Sc_G \leq 1$

$k_{ma1} = 0.328\ (450{,}823)^{-0.227} + 0.298\ (450{,}823)^{-0.127}$

$k_{ma1} = 0.017 + 0.057$

$k_{ma1} = 0.074$

b. $k_{ma5} = 0.431\ Sqig^{-0.129}$ for $Sc_G = 5$

$k_{ma5} = 0.0804$

6. Interpolate between the two curves using calculated Schmidt number, k_m/alpha:

$k_m/\text{alpha} = k_{ma5} - [(5 - Sc_G)/(5 - 1) \times (k_{ma5} - k_{ma1})]$

$k_m/\text{alpha} = 0.0804 - [(5 - 1.734)\ (0.0804 - 0.074)/4]$

$k_m/\text{alpha} = 0.0804 - (0.8165 \times 0.0064)$

$k_m/\text{alpha} = 0.0753$

TABLE A-28. SAMPLE CALCULATIONS FOR OIL/WATER SEPARATORS (Continued)

7. Calculate local mass transfer coefficient, k_m:

k_m = (k_m/alpha) alpha

k_m = 0.0753 x 4.3147 cm/s

k_m = 0.3247 cm/s

8. Calculate concentration of toluene at vapor-liquid interface, C_L:

C_L (gmol/m^3) = P*/RT

C_L (gmol/m^3) = 0.0468 atm/[(82.05 atm-cm^3/gmol°K) (303°K)]

C_L (gmol/m^3) = 1.88 x 10^{-6} gmol/cm^3 x 10^6 cm^3/m^3

C_L = 1.88 gmol/m^3

9. Calculate emission of toluene from an oil-water separator, $(N_a)_T$:

$(N_a)_T$ = k_m x C_L x L x W, mol/s

$(N_a)_T$ = (0.3247 cm/s) (1.88 gmol/m^3) (1371 cm)

(762 cm) (m^3/10^6cm^3)

$(N_a)_T$ = 0.638 gmol/s

$(N_a)_T$ = 0.638 gmol/s (3600 s/hr) (7200 hr/yr)

(92.0 g/gmol) (10^{-6} Mg/g)

$(N_a)_T$ = 1522 Mg/yr

10. Calculate fraction emitted of toluene from an oil-water separator, fe:

$(fe)_T$ = $(N_a)_T/F_i$

$(fe)_T$ = (0.638 gmol/s) (92.0 g/gmol)/(126.18 g/s)

$(fe)_T$ = 0.465

used to estimate emissions based on this theory is presented in Table A-29. According to this theory, the mass transfer rate of each organic compound present in the falling wastewater is controlled by the liquid phase resistance. The gas phase resistance is assumed to be negligible due to the degree of convective mass transfer in this phase. Reaeration rates and physical properties of oxygen are used to estimate the liquid phase resistance to mass transfer. The diffusivity in water of each organic compound are compared to the diffusivity in water of oxygen. The rate of emissions of each organic compounds are then based on this ratio as well as the height of the weir.

A typical weir configuration is shown in Figure A-10. Emission estimates for the weir are based on the typical design dimensions shown in the figure. Table A-30 presents the fraction emitted for five example organic compounds. Each of the five organic compounds are assumed to be present at a concentration of 10 ppm. The wastewater is assumed to be at ambient temperature for the purposes of the calculation. An example calculation for weirs is shown in Table A-31.

A.2 FRACTION EMITTED DURING COLLECTION AND TREATMENT (Fe)

Three example waste stream systems were developed to evaluate potential VOC emissions from different collection and treatment scenarios. Schematics of these example collection and treatment systems are shown in Figures A-11, A-12, and A-13. The individual components discussed in Section A.1 were used to develop the three example schematics. The emission factors (fe) presented for each collection and treatment system component are used in a generalized procedure to estimate VOC emissions from each wastewater stream. The procedure, in estimating the fraction emitted from a component in the system, accounts for the cumulative fraction emitted prior to that component. For example, if the overall emission factor from a component is 0.2 and the cumulative fraction emitted prior to that component is 0.4, then the actual fraction emitted is 0.12 [0.2(1 - 0.4)]; the emission factor, 0.2, is applied only to the fraction of the original organic compounds reaching the component. The fraction of the original organic compounds passing on in the water to the next component is 0.48 (1 -0.4 - 0.12). The emission factors, or fractions emitted, for the individual collection and treatment system components discussed in Section A.1 are summarized in Table A-32.

TABLE A-29. ESTIMATION TECHNIQUE FOR WEIRS

Volatilization - Reaeration Theory

$$C_o = C_i \exp(-K_D(i))$$

where

$$K_D(i) = K_D(O_2)[(D_w)_i/D_{O2,w}]^{0.75}$$

I. Calculate $K_D(O_2)$

$$K_D(O_2) = 0.16h$$

II. Calculate $K_D(i)$

$$K_D(i) = K_D(O_2)[(D_w)_i/D_{O2,w}]^{0.75}$$

III. Calculate fraction emitted, f_e

$$C_o/C_i = \exp(-K_D(i))$$

$$f_e = 1 - (C_o/C_i)$$

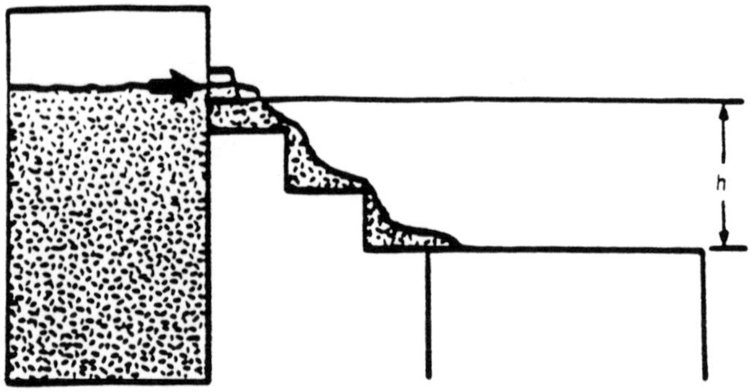

Design Parameter	Range	Typical Design
Weir height, h (m)	0.9 - 2.7	1.8

Figure A-10. Typical weir configuration.

TABLE A-30. EMISSION ESTIMATES FOR WEIRS

Compound	Henry's Law Constant H, atm - m^3/gmol (25°C)	Fraction Emitted, Fe
1,3 Butadiene	1.42E-01	2.9E-01
Toluene	6.68E-03	2.5E-01
Naphthalene	1.18E-03	2.3E-01
1-Butanol	8.90E-06	2.6E-01
Phenol	4.54E-07	2.6E-01

TABLE A-31. SAMPLE CALCULATION FOR WEIRS

<u>Volatilization - Reaeration Theory</u>

$$C_o = C_i \exp(-K_D(i))$$

where

$$K_D(i) = K_D(O_2)[(D_w)_i/D_{O2,w}]^{0.75}$$

and

$$K_D(O_2) = 0.16h$$

where

C_o = effluent concentration of toluene

C_i = influent concentration of toluene

h = 4 ft

T = 25°C

$(D_w)_i = 8.60 \times 10^{-6}\ cm^2/s$

$D_{O2,w} = 2.50 \times 10^{-5}\ cm^2/s$

First, calcualte $K_D(O_2)$

$$K_D(O_2) = 0.16h$$

$$K_D(O_2) = 0.16\ (4) = 0.64$$

(CONTINUED)

TABLE A-31. (CONTINUED)

Then, calculate $K_D(i)$

$$K_D(i) = K_D(O_2)[(D_w)_i/D_{O2,w}]^{0.75}$$

$$K_D(i) = 0.64\ [(8.60 \times 10^{-6})/(2.50 \times 10^{-5})]^{0.75}$$

$$K_D(i) = 0.287$$

Now, calculate fraction emitted

$$C_o = C_i \exp(-K_D(i))$$

rearrange

$$C_o/C_i = \exp(-K_D(i))$$

so,

$$C_o/C_i = \exp(-0.287)$$

$$C_o/C_i = 0.750$$

and, the fraction emitted is:

$$(f_e) = 1 - (C_o/C_i) = 1 - 0.750 = 0.250$$

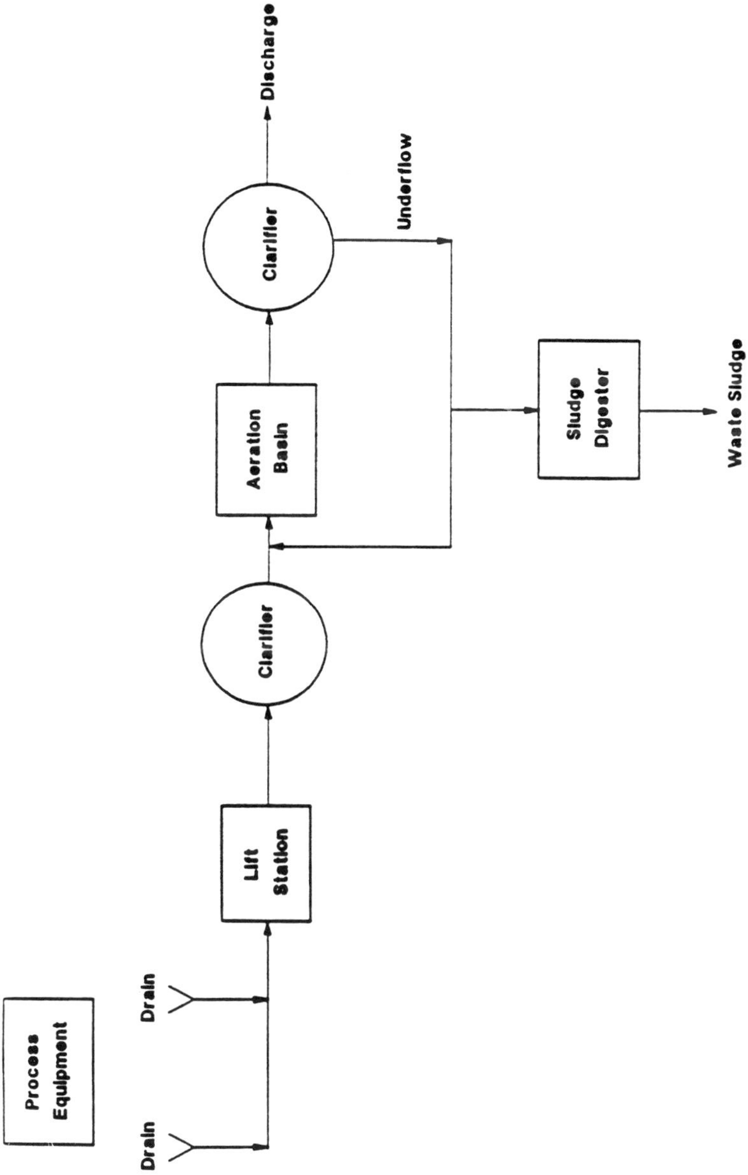

Figure A-11. Example Waste Stream Schematic I.

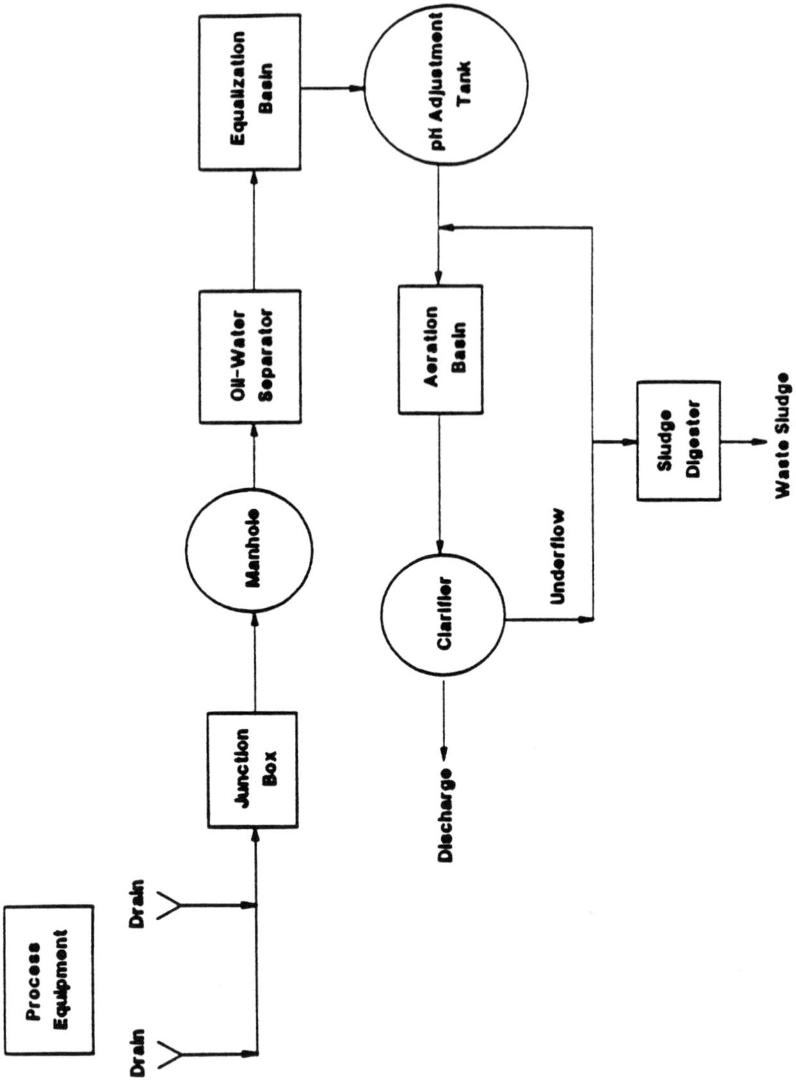

Figure A-12. Example Waste Stream Schematic II.

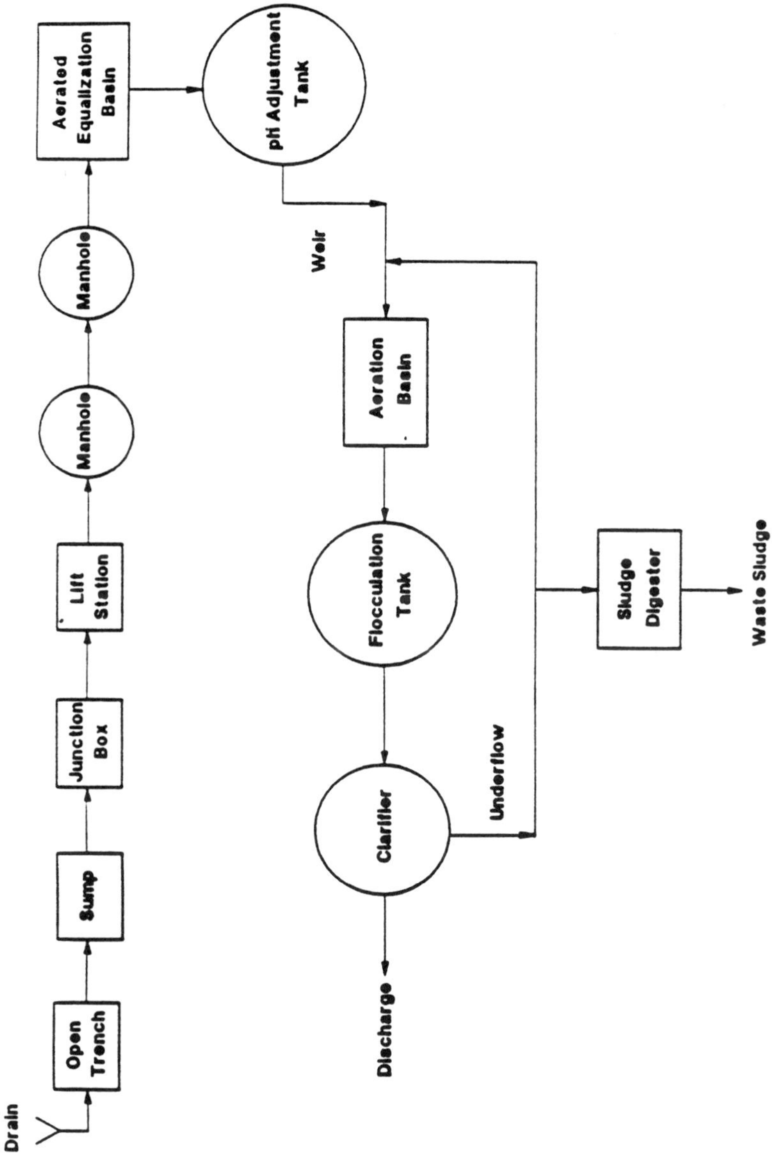

Figure A-13. Example Waste Stream Schematic III.

TABLE A-32. SUMMARY OF EMISSION FACTORS FOR COLLECTION AND TREATMENT SYSTEM COMPONENTS

Component	Emission Factor (Fe)					
	1,3-Butadiene	Toluene	Naphthalene	1-Butanol	Phenol	Overall VOC
Drain	5.7E-01	6.1E-02	1.1E-02	8.7E-05	4.4E-06	1.3E-01
Manhole	1.5E-01	8.2E-03	1.5E-03	1.1E-05	5.7E-07	3.2E-02
Junction Box	1.2E-01	9.8E-02	6.8E-02	1.9E-03	1.1E-04	5.7E-02
Lift Station	3.6E-01	2.9E-01	1.8E-01	3.6E-03	2.0E-04	1.7E-01
Sump	5.6E-03	5.0E-03	4.7E-03	2.1E-03	1.9E-04	3.5E-03
Open Trench	5.9E-02	4.5E-02	2.5E-02	4.1E-04	2.1E-05	2.6E-02
Equalization Basin						
(Non-Aerated)	4.3E-01	4.0E-01	3.8E-01	1.8E-01	1.6E-02	2.8E-01
(Aerated)	1.0E+00	9.9E-01	9.9E-01	6.1E-01	7.7E-02	7.3E-01
Clarifier	2.8E-02	2.5E-02	2.4E-02	7.8E-03	5.6E-04	1.7E-02
Biobasin						
(Aerated)	2.9E-01	1.2E-01	1.7E-01	4.1E-02	1.9E-05	1.2E-01
(Non-Aerated)	2.2E-01	6.0E-02	1.2E-01	3.0E-01	3.9E-04	1.4E-01
Treatment Tank	1.0E-02	9.2E-03	8.6E-03	2.9E-03	2.1E-04	6.2E-03
Oil-Water Separator	1.0E+00	4.7E-01	9.0E-02	6.9E-04	3.7E-05	3.1E-01
Weir	2.9E-01	2.5E-01	2.3E-01	2.6E-01	2.6E-01	2.6E-01

The fractions of the five example organic compounds emitted from each example waste stream system are listed in Table A-33, A-34, and A-35. As seen in the tables, the overall fraction emitted for each organic compound is the highest for Example Waste Stream Schematic III. For example, for toluene, the overall fraction emitted was 0.47, 0.74, and 1.0 from Example Waste Stream Schematics I, II, and III, respectively. For phenol, the compound with the lowest Henry's Law constant, the cumulative fraction emitted remained relatively low; 7.9×10^{-4}, 1.7×10^{-2}, and 3.2×10^{-1} for Example Waste Stream Schematics I, II, and III, respectively.

A.3 EMISSIONS REDUCTION

Emission reduction measures the efficiency of removal for a control technique. Removal efficiencies for the model steam stripper are presented in Chapter 4, while the fraction emitted for each example waste stream is summarized in Section A.2. For steam stripping of wastewaters containing organic compounds, the emission reduction is dependent on the mass of organics in the wastewater, the removal efficiency of the control technique, and the fraction of organic compounds that would be emitted from the wastewater.

Tables A-36 through A-38 summarize the emission reduction as well as the uncontrolled and controlled emissions for Example Waste Stream Schematics I, II, and III, respectively. The following sample calculations for Example Waste Stream Schematic I at 300 lpm illustrate the method used to estimate emission reductions.

Initial Organic Compound Loading in Wastewater:

$$300 \text{ l/min} * 500 \text{ mg/l} * 1 \text{ g}/10^3\text{mg} * 1 \text{ Mg}/10^6\text{g} * 60 \text{ min/hr} * 24 \text{ hr/day} * 365 \text{ days/yr} = 79 \text{ Mg/yr per compound}$$

Uncontrolled Emissions:

ex. Toluene - Fe x Initial Loading

$$0.47 * 79 = 37 \text{ Mg/yr}$$

TABLE A-33. CUMULATIVE FRACTION EMITTED DURING COLLECTION AND TREATMENT
EXAMPLE WASTE STREAM SCHEMATIC I

Component	Cumulative Fraction Emitted					
	1,3-Butadiene	Toluene	Naphthalene	1-Butanol	Phenol	Overall
Drain	5.7E-01	6.1E-02	1.1E-02	8.7E-05	4.4E-06	0.13
Drain	8.2E-01	1.2E-01	2.3E-02	1.7E-04	8.9E-06	0.19
Lift Station	8.8E-01	3.8E-01	2.0E-01	3.8E-03	2.1E-04	0.29
Clarifier	8.9E-01	3.9E-01	2.1E-01	1.2E-02	7.7E-04	0.30
Aeration Basin	9.2E-01	4.7E-01	3.4E-01	5.2E-02	7.9E-04	0.36
Clarifier	9.2E-01	4.7E-01	3.4E-01	5.2E-02	7.9E-04	0.36

TABLE A-34. CUMULATIVE FRACTION EMITTED DURING COLLECTION AND TREATMENT FOR EXAMPLE WASTE STREAM SCHEMATIC II

Component	Cumulative Fraction Emitted					Overall
	1,3-Butadiene	Toluene	Naphthalene	1-Butanol	Phenol	
Drain	5.7E-01	6.1E-02	1.1E-02	8.7E-05	4.4E-06	0.13
Drain	8.2E-01	1.2E-01	2.3E-02	1.7E-04	8.9E-06	0.19
Junction Box	8.4E-01	2.1E-01	8.9E-02	2.1E-03	1.2E-04	0.27
Manhole	8.6E-01	2.1E-01	9.1E-02	2.1E-03	1.2E-04	0.23
Oil-Water Separator	1.0E+00	5.8E-01	1.7E-01	2.8E-03	1.6E-04	0.35
Non-aerated Equalization Basin	1.0E+00	7.2E-01	2.7E-01	1.8E-01	1.6E-02	0.44
pH Adjustment Tank	1.0E+00	7.2E-01	2.7E-01	1.9E-01	1.7E-02	0.44
Aeration Basin	1.0E+00	7.4E-01	3.0E-01	2.2E-01	1.7E-02	0.45
Clarifier	1.0E+00	7.4E-01	3.0E-01	2.2E-01	1.7E-02	0.45

TABLE A-35. CUMULATIVE FRACTION EMITTED DURING COLLECTION AND TREATMENT FOR EXAMPLE WASTE STREAM SCHEMATIC III

Component	Cumulative Fraction Emitted					
	1,3-Butadiene	Toluene	Naphthalene	1-Butanol	Phenol	Overall
Drain	5.7E-01	6.1E-02	1.1E-02	8.7E-05	4.4E-06	0.13
Open Trench	6.0E-01	1.0E-01	3.6E-02	5.0E-04	2.6E-05	0.15
Sump	6.0E-01	1.1E-01	4.1E-02	2.6E-03	2.2E-04	0.15
Junction Box	6.5E-01	2.0E-01	1.1E-01	4.5E-03	3.3E-04	0.19
Lift Station	7.7E-01	4.3E-01	2.6E-01	8.1E-03	5.3E-04	0.30
Manhole	8.1E-01	4.4E-01	2.7E-01	8.1E-03	5.3E-04	0.30
Manhole	8.4E-01	4.4E-01	2.7E-01	8.1E-03	5.3E-04	0.31
Aerated Equalization Basin	1.0E+00	1.0E+00	9.9E-01	6.1E-01	7.7E-02	0.73
pH Adjustment Tank	1.0E+00	1.0E+00	9.9E-01	6.1E-01	7.8E-02	0.74
Weir	1.0E+00	1.0E+00	9.9E-01	7.1E-01	3.2E-01	0.80
Aeration Basin	1.0E+00	1.0E+00	9.9E-01	7.2E-01	3.2E-01	0.81
Flocculation Tank	1.0E+00	1.0E+00	9.9E-01	7.2E-01	3.2E-01	0.81
Clarifier	1.0E+00	1.0E+00	9.9E-01	7.2E-01	3.2E-01	0.81

TABLE A-36 EMISSION REDUCTION FOR EXAMPLE WASTE STREAM SCHEMATIC I

Compound	Fraction Removed by Stripper	Uncontrolled Loading (Mg/yr)[a]	Controlled Loading (Mg/yr)	Fraction Emitted[b]	Uncontrolled Emissions (Mg/yr)	Controlled Emissions (Mg/yr)	Percent Emission Reduction (%)
Toluene	1.0	79	0.000	0.47	37	0.000	100%
Naphthalene	1.0	79	0.000	0.34	27	0.000	100%
1,3-Butadiene	1.0	79	0.000	0.92	72	0.000	100%
Phenol	0.022	79	77	0.001	0.06	0.061	2.4%
Butanol	0.30	79	52	0.052	4.1	2.7	0%
Total VOC		400	130		140	2.8	98%

[a]Calculated at 300 lpm, 500 ppm per compound, 365 days/yr.

[b]Table A-33.

TABLE A-37. EMISSION REDUCTION FOR EXAMPLE WASTE STREAM SCHEMATIC II

Compound	Fraction Removed by Stripper	Uncontrolled Loading (Mg/yr)[a]	Controlled Loading (Mg/yr)	Fraction Emitted[b]	Uncontrolled Emissions (Mg/yr)	Controlled Emissions (Mg/yr)	Percent Emission Reduction (%)
Toluene	1.0	79	0.00	0.74	59	0.00	100%
Naphthalene	1.0	79	0.00	0.30	23	0.00	100%
1,3-Butadiene	1.0	79	0.00	1.0	79	0.00	100%
Phenol	0.022	79	76	0.017	1.3	1.3	2.2%
Butanol	0.30	79	52	0.22	17	11	30%
Total VOC		400	130		180	13	93%

[a]Calculated at 300 lpm, 500 ppm per compound, 365 days/yr.

[b]Table A-34.

TABLE A-38. EMISSION REDUCTION FOR EXAMPLE WASTE STREAM SCHEMATIC III

Compound	Fraction Removed by Stripper	Uncontrolled Loading (Mg/yr)[a]	Controlled Loading (Mg/yr)	Fraction Emitted[b]	Uncontrolled Emissions (Mg/yr)	Controlled Emissions (Mg/yr)	Percent Emission Reduction (%)
Toluene	1.0	79	0.00	1.0	79	0.00	100%
Naphthalene	1.0	79	0.00	0.99	79	0.00	100%
1,3-Butadiene	1.0	79	0.00	1.0	79	0.00	100%
Phenol	0.022	79	77	0.32	25	24	2.2%
Butanol	0.30	79	52	0.72	57	38	30%
Total VOC		400	130		320	60	80%

[a]Calculated at 300 lpm, 500 ppm per compound, 365 days/yr.

[b]Table A-35.

Controlled Emissions:

ex. Toluene -

Uncontrolled Loading * (1 - fraction removed (fr) by stripper) * fe

79 * (1 - 1) * 0.47 = 0

Total % of Emission Reduction (ER) for VOC

$$\frac{\text{Uncontrolled Emissions - Controlled Emissions}}{\text{Uncontrolled Emissions}}$$

140 - 2.8 / 140 * 100 = 98%

A.4. REFERENCES

1. Trip Report. D. J. Herndon, Radian Corporation, to file. 11 p. Report of May 4, 1987, visit to Borden Chemical Company.

2. Trip Report. D. J. Herndon, Radian Corporation, to file. 4 p. Report May 5, 1987, visit to Union Carbide Corporation.

3. Trip Report. D. J. Herndon, Radian Corporation, to file. 13 p. Report May 8, 1987, visit to Dow Chemical Company.

4. Trip Report. D. J. Herndon and S. K. Buchanan, Radian Corporation, to file. 10 p. Report of May 22, 1987, visit to Rhone-Poulenc Agricultural (RP-Ag) Company.

5. Trip Report. D. J. Herndon, Radian Corporation, to file. 5 p. Report of May 6, 1987, visit to PPG Industries.

6. Trip Report. R. H. Howle and M. A. Vancil, Radian Corporation, to file. 7 p. Report of May 12, 1987, visit to Allied Fibers.

7. Trip Report. D. J. Herndon, Radian Corporation, to file. 7 p. Report of May 7, 1987, visit to Rohm & Haas Texas, Inc.

8. Trip Report. D. J. Herndon and S. K. Buchanan, Radian Corporation, to file. 7 p. Report of May 21, 1987, visit to Mobay Chemical Company.

9. Trip Report. D. J. Herndon, Radian Corporation, to file. 6 p. Report of May 29, 1987, visit to General Electric.

10. Trip Report. D. J. Herndon and S. K. Buchanan, Radian Corporation, to file. 7 p. Report of September 3, 1987, visit to E. I. duPont de Nemours Company (Deepwater, New Jersey).

11. Trip Report. D. J. Herndon and S. K. Buchanan, Radian Corporation, to file. 6 p. Report of September 2, 1987, visit to Fritzsche Dodge & Olcott.

12. Trip Report. D. J. Herndon and S. K. Buchanan, Radian Corporation, to file. 6 p. Report of September 1, 1987, visit to Vista Chemical Company.

13. Trip Report. D. J. Herndon and S. K. Buchanan, Radian Corporation, to file. 5 p. Report of September 3, 1987, visit to CIBA-GEIGY Corporation.

14. Trip Report. D. J. Herndon, Radian Corporation, to file. 4 p. Report of March 1, 1988, visit to Morflex Chemical Company, Incorporated.

15. Trip Report. D. J. Herndon, Radian Corporation, to file. 4 p. Report of March 8, 1988, visit to Burroughs Wellcome.

16. Trip Report. D. J. Herndon, Radian Corporation, to file. 5 p. Report of September 9, 1987, visit to Monsanto Company.

17. Trip Report. D. J. Herndon, Radian Corporation, to file. 5 p. Report of March 31, 1988, visit to Shell Deer Park Manufacturing Complex.

18. Trip report. D. J. Herndon, Radian Corporation, to file. 4 p. Report of February 19, 1988, visit to Ethyl Corporation.

19. Trip Report. D. J. Herndon, Radian Corporation, to file. 5 p. Report of March 30, 1988, visit to Hoechst-Celanese.

20. Trip report. D. J. Herndon, Radian Corporation, to file. 4 p. Report of March 31, 1988, visit to E. I. duPont deNemours and Company (LaPorte, Texas).

21. Trip Report. J. A. Elliott, Radian Corporation, to file. 4 p. Report of September 18, 1989, visit to Glaxo, Incorporated.

22. Office of Air Quality, Planning and Standards. U. S. Environmental Protection Agency. Hazardous Waste Treatment, Storage, and Disposal Facilities (TSDF) - Air Emission Models, Draft Report. April 1989.

23. Metcalf and Eddy, Inc. Wastewater Engineering. New York, McGraw-Hill. 1972. p. 519.

24. Process Design Manual for Suspended Solids Removal, U. S. Environmental Protection Agency, Technology Transfer, January 1975. EPA-625/1-75-003a.

25. Metcalf and Eddy, Inc. Wastewater Engineering. New York, McGraw-Hill. 1972. p. 525.

26. Liang, S. F. Hydrocarbon Losses by Atmospheric Evaporation from Open Separators, Technical Progress Report, BRC-Corp-24-73-F. (Shell Models)

27. Office of Air Quality Planning and Standards, U. S. Environmental Protection Agency. Research Triangle Park, NC. VOC (Volatile Organic Compound) Emissions from Petroleum Refinery Wastewater System - Background Information for Proposed Standards, Draft EIS. EPA-450/3-85-001a (NTIS PB87-190335), February 1985.

Appendix B—Estimation of Air Emission Factors from Airflow in Wastewater Collection Systems

B.1 INTRODUCTION

The purpose of the theoretical analysis of wastewater collection systems is to obtain air emission factors that can be used to estimate the release of volatiles into the atmosphere. As a volatile waste constituent is discharged into a collection system, it can be emitted into the atmosphere through air flowing through the collection system. Air can enter and leave a collection system by openings in drains, open channels, channels with grates, openings in manhole covers, junction boxes, sumps, and other openings. Estimation of the flow of air in a collection system unit (drain, manhole) relative to the flow of wastewater flowing under the collection system unit permits an estimation of the fraction of the volatile constituent lost to the atmosphere as it passes under the unit.

The assumptions that were made to characterize chemical sewer designs include the following:

- The design depth in the drain channel is assumed to be half full.
- The flow in the channel for estimating fractional emissions is assumed to be 80 percent of design depth. (Lower depths result in higher emissions.)
- The air exiting the system is assumed to be at equilibrium with the volatiles in the channels.
- A typical wind is assumed to be 3.5 mph.

The emission factors for the collection units are sensitive to the magnitude of the flow rates in the channels. The loss of volatiles in the channels could be less than the equilibrium amount if the rate of mass transfer from the bulk of the wastewater to the air were to limit the rate of air emissions. This mass transfer rate is expected to be sensitive to the depth in the channel, with equilibrium not achieved for high flows of air across deep channels. For the case of channel depths at a fraction of the design depths and relatively low air rates (manhole covers and enclosed collection systems), the assumption of equilibrium is expected to be appropriate.

Since the air emission factors are sensitive to environmental factors such as temperature, humidity, and wind pressure, the measured air emissions from wastewater collection systems are expected to be variable.

B.2 DISCUSSION OF THE USE OF COLLECTION SYSTEM EMISSION FACTORS

The emission factors developed in this report are expressed in terms of fraction of material in the sewer main emitted per unit. When the path of the waste placed in the collection system is specified, the amount of material remaining in the original waste stream is recalculated each time the waste flows under a potential emission source (drain connection, manhole, lift station, sump, etc.):

Emissions from unit = amount present x unit emission factor

New amount present = amount present - emissions from unit.

The following example illustrates how the toluene emissions from a waste discharge into a collection system can be estimated. The waste flows into an open trench drain. Forty feet downstream, additional waste flows into the trench for an additional 20 ft. The flow in the trench discharges into a drain. The subsurface channel in the sewer has an additional drain connection and a manhole before discharge into a covered sump with a vent.

Unit	Emission factor	Amount present, g	Emissions, g
Open trench drain (40 ft)	0.045	100	4.5
Open trench drain (20 ft)	0.022	95.5	2.1
Drain	0.08	93.4	7.5
Drain connection	0.08	85.9	6.7
Manhole at junction	0.0083	79.2	0.66
Covered sump with vent	0.11	78.5	8.6
Overall collection units	0.30	70	30

This application of the unit emission factors to a wastewater collection system for toluene wastes indicates that a substantial fraction of the original toluene in the waste can be lost due to airflows in the collection system. Another way of interpreting these data is that for every 70 g of toluene that enter the wastewater treatment plant, 30 g are emitted in the collection system before reaching the wastewater treatment plant (43 percent).

These emission factors for wastewater collection systems are not expected to be applicable for all systems. They are for a wastewater collection system designed to aerate the wastewater, either for safety, for corrosion reduction, or for odor control. There are a number of equipment changes that can reduce the air emissions to levels much lower than can the system presented here. Emissions can be reduced by using covers for sumps, manhole covers with fewer and smaller openings, seals on drain openings, or solid metal covers for trenches; by purging the system with excess water; and by other methods. Emissions from the collection systems presented here can be increased also by high winds, discharge of steam into the sewer, open sumps, open junctions, complex collection systems with many units (potential emission sources) before discharge, and other factors.

It is possible that the emission factors presented in this report will be modified in the future. Factors that could be used to improve the accuracy of the emission factors include considering mass transfer at the liquid gas interface and using Monte Carlo simulations of collection system characteristics.

B.3 METHODS AND RESULTS

B.3.1 Overview

Air emissions factors are presented for induced airflow in sewer systems accepting hazardous aqueous waste. The major sources of induced airflow into and out of a sewer system are process drains, manholes, and junction boxes. The emission factors are generated for five different organic compounds: 1,3-butadiene, toluene, naphthalene, 1-butanol, and phenol.

Ten cases for induced airflow in sewers are illustrated in Figure B-1. Cases A1, A2, and A3 illustrate potential airflows from process drains.

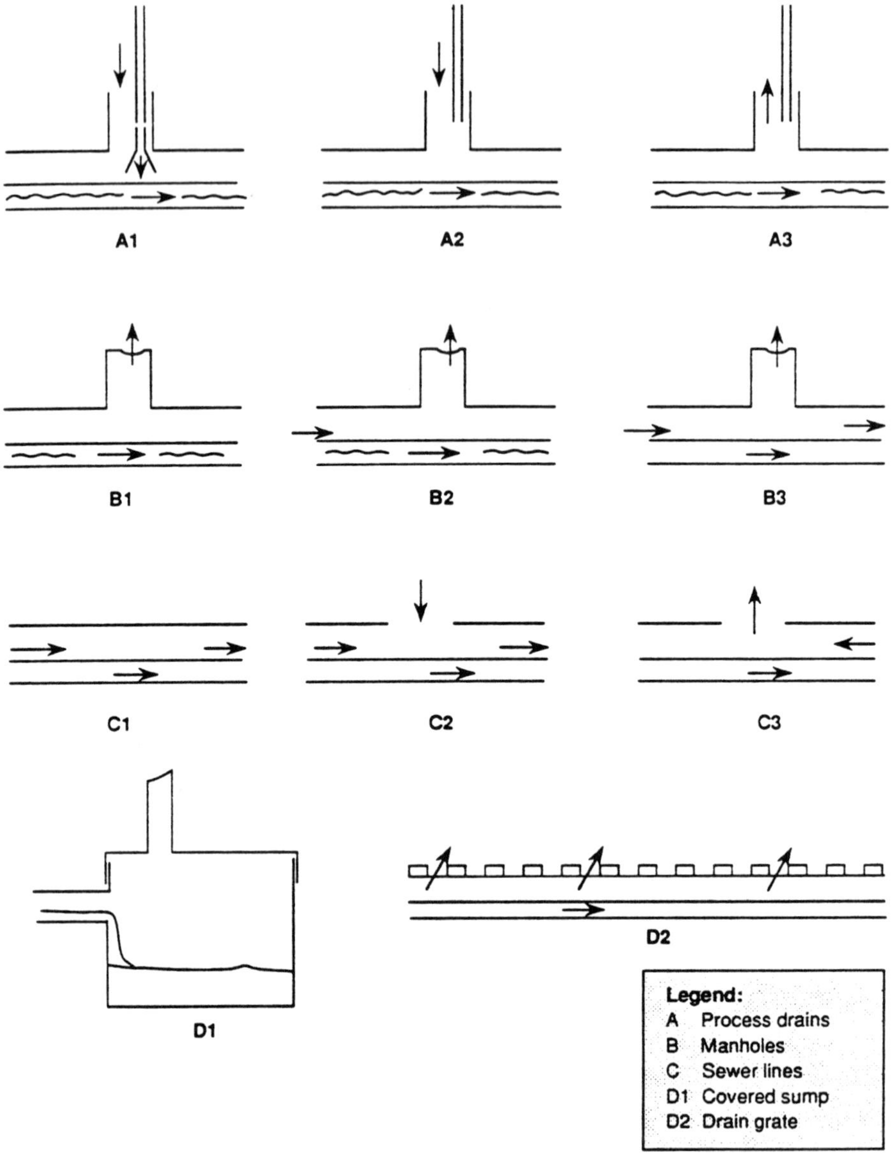

Figure B-1. Simplified flow diagrams.

Cases B1, B2, and B3 illustrate air emissions from manholes. Cases C1, C2, and C3 illustrate airflow out of sewer lines. Case D1 represents emissions from a covered sump with an open vent, and Case D2 illustrates airflow out of drain grates. The following brief paragraphs describe some of the assumptions used in estimating the induced flow of air:

- Case A1 estimates airflow into a drain annulus induced by water flow. The air drawn in will escape somewhere and be in equilibrium with the water at that point.

- Case A2 estimates airflow into a sewer through a drain annulus. No water is flowing into the drain. The air comes to equilibrium with the water flowing in the sewer and escapes at some point upstream or downstream of the drain.

- Case A3 estimates airflow from saturated air rising from a drain annulus due to a density difference between the air in the sewer and the ambient air. No water is flowing through the drain. The air is drawn in at a point upstream or downstream of the drain and reaches thermal and chemical equilibrium with the wastewater flowing in the sewer by the time it reaches the drain.

- Case B1 estimates airflow from manhole cover vents caused by a density difference between air in the sewer and the ambient air. The air flowing out of the vents is in thermal and chemical equilibrium with the water flowing in the sewer at that point.

- Case B2 estimates airflow through manhole cover vents induced by wind blowing in the upstream end of a sewer that is blocked off after the manhole. The air is in equilibrium with the water in the manhole.

- Case B3 estimates the airflow from manhole cover vents induced by wind blowing in one end of a sewer and flowing past the manhole to some point downwind. The air is in equilibrium with the water in the sewer at the manhole. No drains or vents are in the line between the upwind sewer end and the manhole.

- Case C1 estimates the airflow induced by wind blowing in one end of a sewer and out another. The air is in equilibrium with water at the downwind end of the sewer.

- Case C2 estimates the airflow into the sewer from a junction box induced by water flow through the junction box. This air escapes somewhere (e.g., the next junction box downstream) in equilibrium with the water flowing through at that point.

- Case C3 estimates airflow from the discharge end of a partially filled sewer resulting from density differences between the ambient air and the warm humid air in equilibrium with the wastewater.
- Case D1 estimates the airflow induced through a stack on an enclosed sump. Air is in equilibrium with the wastewater and is drawn into the system at some point upstream or downstream of the sump.
- Case D2 estimates airflow from an open trench based upon mass transfer in the rapid flowing water.

Tables B-1 through B-7 describe the estimated fraction of the volatile organic emitted from the three components of the sewer investigated for five different compounds that differ in volatility.

The airflow induced by the wind is sensitive to the geometry of the source, the direction of flow of the wind, and the velocity of the wind. Because of the large numbers of significant factors that could conceivably influence the rate of emissions due to wind, the emission estimates are presented as a range, with zero as the lower bound of the range and a combination of values from the three cases as the upper range. The choice of a specific value to be used for estimating emission factors from induced airflow in the sewer component is also presented in Tables B-1 to B-7. In some cases, the effects of the various mechanisms for airflow can be additive, but in some cases the effects would tend to cancel each other.

B.3.2 Description of Case A1 Calculations

Case A1 considers airflow into a drain induced by wastewater discharged to the sewer through a pipe inserted in the drain. The air is assumed to be drawn into the annulus with a velocity equal to that of the flowing water at the air/water interface. The velocity of the induced air is assumed to decrease to zero at the wall of the drain. The assumed air velocity profile has not been experimentally confirmed. The air drawn into the drain is assumed to escape at some other point in the system after coming to equilibrium with the wastewater. In relatively tight systems or systems with long runs of sewer between openings, the resistance to airflow will inhibit this mechanism of air induction.

TABLE B-1. EMISSION ESTIMATES FOR DILUTE AQUEOUS 1,3-BUTADIENE SOLUTIONS FLOWING THROUGH SEWER NETWORKS[a] (FRACTION EMITTED)

	Drains (A)	Manholes (B)	Sewers (C)
Case 1	0.63	0.087	0.95
Case 2	0.73	0.21	0.79
Case 3	0.54	0.147	0.56
Typical value	0.63	0.15	0.77

[a]Case A1 is Unit A with Case 1 conditions.

TABLE B-2. EMISSION ESTIMATES FOR DILUTE AQUEOUS TOLUENE SOLUTIONS FLOWING THROUGH SEWER NETWORKS (FRACTION EMITTED)

	Drains (A)	Manholes (B)	Sewers (C)
Case 1	0.073	0.0045	0.48
Case 2	0.113	0.0123	0.148
Case 3	0.053	0.008	0.057
Typical value	0.08	0.0083	0.23

TABLE B-3. EMISSION ESTIMATES FOR DILUTE AQUEOUS NAPHTHALENE SOLUTIONS FLOWING THROUGH SEWER NETWORKS (FRACTION EMITTED)

	Drains (A)	Manholes (B)	Sewers (C)
Case 1	0.014	0.0008	0.14
Case 2	0.022	0.0022	0.030
Case 3	0.0098	0.0014	0.02
Typical value	0.015	0.0015	0.06

TABLE B-4. EMISSION ESTIMATES FOR DILUTE AQUEOUS 1-BUTANOL SOLUTIONS FLOWING THROUGH SEWER NETWORKS (FRACTION EMITTED)

	Drains (A)	Manholes (B)	Sewers (C)
Case 1	0.0001	0.000006	0.00123
Case 2	0.00017	0.000017	0.00023
Case 3	0.00007	0.000011	0.00008
Typical value	0.00012	0.00001	0.0005

TABLE B-5. EMISSION ESTIMATES FOR DILUTE AQUEOUS PHENOL SOLUTIONS FLOWING THROUGH SEWER NETWORKS (FRACTION EMITTED)

	Drains (A)	Manholes (B)	Sewers (C)
Case 1	0.0000053	$3 \cdot 10^{-7}$	0.000063
Case 2	0.0000086	$8.5 \cdot 10^{-7}$	0.000012
Case 3	0.0000038	$5.5 \cdot 10^{-7}$	0.0000041
Typical value	0.000006	$6 \cdot 10^{-7}$	0.000026

TABLE B-6. EMISSION ESTIMATES FROM AN OPEN-TRENCH SECTION IN A WASTEWATER COLLECTION NETWORK

Compound	Partition coefficient (Y/X)	Fraction emitted to air
1,3-butadiene	7,900	0.059
Toluene	371	0.045
Naphthalene	65.6	0.025
Butanol	0.494	0.0004
Phenol	0.0252	0.0002

TABLE B-7. PARTITION COEFFICIENTS OF COMPOUNDS USED IN EMISSION ESTIMATES[a]

Compound	K
1,3-Butadiene	4000
Toluene	371
Naphthalene	65.5
1-Butanol	0.494
Phenol	0.0252

[a]Mole fraction in gas phase/mole fraction is aqueous waste.

The calculation requires the following inputs: flow rate of wastewater, ratio of wastewater pipe area to drain pipe area, partition coefficient applicable to the pollutant of interest at the wastewater temperature, concentration of wastewater stream, and temperature of the ambient air. The molar air density is calculated at the ambient temperature based on the ideal gas law assuming an ambient pressure of one atmosphere. The influent flow rate of volatile organics is calculated from the mass flow rate of wastewater and the mass fraction of volatile organics in the wastewater. The influent air linear flow rate is calculated as one-fourth the linear wastewater flow rate based on the assumed airflow profile. This is converted to a molar airflow rate by multiplying by the area ratio (drain pipe area to wastewater pipe area) and the molar density of air. The fraction emitted is calculated by multiplying the dimensionless partition coefficient by the ratio of molar flows of air to the total molar flow of air and water.

Specify: area ratio of sewer segment (Arr), dimensionless
partition coefficient, K, dimensionless
air temperature, T_a, K

Assume: water density, 0.0555 mol/cm^3
air velocity profile (as described above)
air density by ideal gas law, 0.0121/T mol/cm^3

Calculate: F = fraction emitted:

$$F = \frac{(Arr)(0.25)(0.0121/T_a)K}{(Arr)(0.25)(0.0121/T_a)K + 0.0555}$$

arr	K	T_a	F
4	371	298	0.21
4	371	273	0.23
4	0.5	298	0.00037
4	0.5	273	0.00040
13.7	371	298	0.48
13.7	371	273	0.50
13.7	0.5	298	0.00125
13.7	0.5	273	0.00137

Note that, within the limits of the assumption, a smaller wastewater pipe flowing at an equivalent volumetric flow rate will induce a greater airflow (and cause greater emissions) due to its higher linear velocity.

Note also that slightly greater emissions will occur on cooler days because more moles of denser ambient air will be drawn in (it is assumed that this air will come to thermal equilibrium with the wastewater before it escapes from the system).

B.3.3 Description of Case A2 Calculations

Case A2 considers airflow into a drain and through the sewer. No water is flowing down the drain. The pressure creating the airflow is due to changes in wind velocity. The air pressure is estimated from the maximum pressure obtained from wind flowing at 160 cm/s (3.5 mph) with the pitot tube pointed into the wind. The drain would not normally be oriented into the wind, but wind flow patterns and pressures are expected to be influenced by the location of the drain relative to wind, buildings, sumps, etc.

The air flowing into the drain is assumed to escape at some other point in the system after coming to equilibrium with the wastewater. The frictional drag on the drain and in the headspace of the sewer will determine the flow of air in response to the pressure exerted by the wind.

The maximum pressure exerted by the wind is calculated based on a solution of the Bernoulli equation:

$$dP = \nu^2 \ \rho/(2\ g_c) \quad ,$$

where:

dP = calculated pressure, g force/cm^2
ν = wind velocity, 156 cm/s (3.5 mph)
ρ = density of air at 25 °C, 0.0012 g/cm^3
g_c = 980.665 g-cm/gF-s^2.

$$dP = \frac{(156)^2\ (0.0012)}{2\ (980.665)} = 0.015 \text{ g force/cm}^2 \quad .$$

This value of the maximum pressure is equated to the energy of the air velocity in the sewer and the frictional losses in the sewer:

$$dP/\rho = (1 + Ke + 4\ F\ L/D\ Arr^2 + 4\ F\ L2/D2 + K1)\ \nu^2/2g_c \quad ,$$

where:

- dP = pressure, 0.015 g force/cm^2
- ρ = density of air, g/cm^3
- Ke = diameter change coefficient, 0.31
- F = friction factor of air, 0.006
- L = length of sewer, 1,220 cm
- D = equivalent diameter of the headspace in the sewer, 40.4 cm (four times the hydraulic radius)
- Arr = area ratio of sewer segment
- L2 = length of drain, 61 cm x 2 drains = 122 cm
- D2 = diameter of drain, 20.3 cm
- Kl = loss coefficient, 4
- g_c = 980.665 g-cm/gF-s^2.

Solving for V, the velocity of air in the drain is 62 cm/s (122 ft/min). The sectional area of the headspace is 1,830 cm^2, permitting a calculated airflow of 20,000 cm^3/s. The molar density of the air is $4.0 \cdot 10^{-5}$ mol/cm^3); a molar airflow rate is then calculated as ($8.79 \cdot 10^5$ cm^3/s) ($3.9 \cdot 10^{-5}$ mol/cm^3), or 0.81 mol/s. The flow rate of volatile organics in the air at equilibrium with the initial concentration of volatile organics in the water is as follows:

(0.81 mol/s)(371)(0.0005 g toluene/g water)(18 g water/mol) or 2.7 g/s .

The fraction of volatile organics present in the air at equilibrium is independent of concentration (as long as K is a constant). The fraction of volatile organics in the air is the ratio of the mass flow in the air divided by the sum of the mass flow in the air and water:

$$f = 2.7/(21.1 + 2.7) = 0.11 \quad .$$

General assumptions and calculations:

Air temperature	25 °C
Relative humidity	50 percent
Sewer temperature	30 °C
Friction factor for air	0.006

Wind velocity	156 cm/s (3.5 mph)
Radius of sewer	30.48 cm (12 in.)
Depth of liquid in sewer	24.4 cm (9.6 in.)
Headspace hydraulic radius	10.9 cm
Flow of water in sewer	42,000 cm^3/s
Headspace area in sewer	1,830 cm^2
Density of air at 25 °C	0.0012 g/cm^3
K partition coefficient (Y/X)	371
Weight fraction volatile organics in water	0.0005
Flow of volatile organics in sewer water	21.1 g/s
Molar density of air in sewer	0.00004 mol/cm^3.

B.3.4 Description of Case A3 Calculations

Case A3 considers airflow up from a drain induced by density differences between the ambient air outside the manhole and the warm humid air in the sewer. No water is flowing in the drain. The wastewater in the sewer is assumed to be flowing in a direction perpendicular to the airflow through the vents; the air is assumed to be saturated with water and at chemical and thermal equilibrium with the wastewater. In the case considered, the drain is assumed to be 10 cm (4 in.) in diameter and 61 cm (2 ft) long. Frictional losses through both the drain and the sewer are considered, based on a friction factor of 0.06. The height of the "stack" is assumed to be 61 cm (2 ft). This is the vertical distance between the level of the water in the sewer and the drain. Ambient conditions are assumed to be 25 °C and 50 percent relative humidity. The wastewater temperature is assumed to be 30 °C.

The densities of ambient air and warm humid sewer air are calculated, and the density difference across the drain system is calculated as $8.39 \cdot 10^{-5}$ g/cm^3. The maximum pressure from density differences is the product of the density difference and height. This value of the maximum pressure from density differences is equated to the energy of the air velocity in the sewer and the frictional losses in the sewer:

$$dP/\rho = (1 + Ke + 4\ F\ L/D\ Arr^2 + 4\ F\ L2/D2 + K1)\ \nu^2/2g_c \quad ,$$

where:

dP = pressure, 0.00195 g force/cm^2

ρ = density of air, 0.0012 g/cm^3

Ke = diameter change coefficient, 0.31

F = friction factor of air, 0.006

L = length of sewer, 610 cm

D = equivalent diameter of the headspace in the sewer, 43.6 cm (four times the hydraulic radius)

Arr = area ratio of sewer segment, 0.219

L2 = length of drain, 61 cm

D2 = diameter of drain, 20.3 cm

Kl = loss coefficient, 3

g_c = 980.665 g-cm/gF-s^2.

Solving for V, the velocity of air in the sewer is 4.8 cm/s (9.5 ft/min). The sectional area of the headspace is 1,828 cm^2, permitting a calculated airflow of 8,853 cm^3/s. The molar density of the air is $4.0 \cdot 10^{-5}$ mol/cm^3; a molar airflow rate is then calculated as (8,853 cm^3/s)($4.0 \cdot 10^{-5}$ mol/cm^3), or 0.356 mol/s. The flow rate of volatile organics in the air at equilibrium with the initial concentration of volatile organics in the water is as follows:

$$(0.356 \text{ mol/s})(371)(0.0005 \text{ g/g})(18 \text{ g/mol}) \text{ or } 1.19 \text{ g/s} \quad .$$

The fraction of volatile organics present in the air at equilibrium is independent of concentration (as long as K is a constant). The fraction of volatile organics in the air is the ratio of the mass flow in the air divided by the sum of the mass flow in the air and water:

$$f = 1.19/(21.3 + 1.19) = 0.053 \quad .$$

General assumptions and calculations:

Air temperature	25 °C
Relative humidity	50 percent
Sewer temperature	30 °C
Friction factor for air	0.006
Radius of sewer	30.48 cm (12 in.)
Depth of liquid in sewer	24.4 cm (9.6 in.)
Headspace hydraulic radius	10.9 cm
Flow of water in sewer	42,000 cm^3/s
Headspace area in sewer	1,828 cm^2

Density of saturated air at 40 °C	0.00117 g/cm^3
Density of air at 25 °C	0.0012 g/cm^3
K partition coefficient (Y/X)	371
Weight fraction volatile organics in water	0.0005
Flow of volatile organics in sewer water	21.3 g/s
Molar density of air in sewer	0.00004 mol/cm^3

B.3.5 Description of Case B1 Calculations

Case B1 considers airflow from the vents in a manhole cover induced by density differences between the ambient air outside the manhole and the warm humid air in the sewer. The wastewater in the sewer is assumed to be flowing in a direction perpendicular to the airflow through the vents; the air is assumed to be saturated with water and at chemical and thermal equilibrium with the wastewater. In the case considered, the manhole cover is assumed to have four vent holes of 2.5 cm (1 in.) diameter. Frictional losses through the manhole are assumed negligible relative to losses through the manhole cover vents. The height of the "stack" is assumed to be 67 cm (2 ft). This is the vertical distance between the level of the water in the sewer and the manhole cover. Ambient conditions are assumed to be 25 °C and 50 percent relative humidity. The wastewater temperature is assumed to be 30 °C.

The densities of ambient air and warm humid sewer air are calculated, and the density difference across the manhole cover is determined. The gas velocity through the manhole cover vents was then calculated from the density difference using the equation for a sharp edged orifice:

$$\nu = 0.61\ (2\ g_c\ h\ \Delta\rho/\rho)^{0.5} \quad ,$$

where:

ν = linear velocity through the vent hole, cm/s

g_c = gravitational constant, 981 cm/s^2

h = height of manhole, 61 cm (2 ft)

$\Delta\rho$ = density difference of air above and below manhole, $3.2 \cdot 10^{-5}$ g/cm^3

ρ = density of warm humid air, 0.00117 g/cm^3.

(Frictional losses through the thickness of the cover are negligible.)

The air velocity is converted to a volumetric flow rate by multiplying by the cross-sectional area of the vent holes, 20 cm^2 (0.022 ft^2). Based on this airflow, 710 cm^3/s, the wastewater flow in the sewer, and a partition coefficient appropriate for the compound of interest at the wastewater temperature, the fractional emission is calculated. (The wastewater flow is 2,360 mol/s and was calculated from an assumed sewer size, slope, roughness, and an assumed wastewater depth in the sewer.) The fraction emitted is calculated as F = GK/(GK + L):

$$\frac{0.0285 \times 371}{0.0285 \times 371 + 2{,}360} = 0.0045 \quad ,$$

where:

F = fraction emitted through cover vents, dimensionless

G = airflow rate from the cover vents, 0.0285 mol/s

K = 371, air/water partition coefficient for compound of interest at wastewater temperature, dimensionless

L = wastewater flow rate through sewer, 2,360 mol/s.

B.3.6 Description of Case B2 Calculations

Case B2 estimates airflow through manhole cover vents resulting from wind blowing into the upstream end of a sewer. The air flows down the sewer to the manhole where further airflow is obstructed. This might occur where a sewer ends at a pump sump or where a change in pipe size or slope results in a completely filled pipe with no air space. The airflow rate is estimated by calculating the air velocity through the manhole cover vents that would result in a frictional head loss equal to that available from the wind blowing into the upstream end of the sewer. Frictional losses through the sewer, the manhole, and the cover thickness are assumed to be negligible in comparison to losses through the cover vents.

Frictional losses through the cover vents are calculated using an equation for flow through a sharp-edged orifice:

$$\nu = 0.61/(\nu_w^2 \, \rho_a/\rho_s)^{0.5} \quad ,$$

where:

ν = linear velocity through vent cover, cm/s

ν_w = wind velocity, 156 cm/s (3.5 mph)

ρ_a = ambient air density, 0.0012 g/cm^3

ρ_s = density of warm humid air in sewer, 0.00117 g/cm^3.

The manhole cover is assumed to have four vents of 2.5 cm (1 in.) diameter. The wind velocity in the direction of the sewer is assumed to be 156 cm/s (3.5 mph). The factor of 0.61 is an orifice coefficient that will be approximately constant for the range of flows considered.

The molar airflow rate can be calculated from the linear velocity through the cover vents by multiplying by the total area of the four vents, 20 cm^2 (0.022 ft^2), and dividing by the molar density at the warm humid sewer conditions, 0.00004 mol/cm^3. The wastewater flow rate in the sewer is implicitly specified on the basis of assumed sewer depth, diameter, slope, and roughness (2,360 mol/s).

The fractional emission of volatiles is calculated from the molar flow rates of air and water, and a dimensionless partition coefficient appropriate for the compounds of interest at the wastewater temperature:

$$F = \frac{GK}{GK + L} = 0.123 \quad ,$$

where:

G = airflow rate, 0.0793 mol/s

K = 371, dimensionless partition coefficient.

L = water flow rate, 2,360 mol/s.

B.3.7 Description of Case B3 Calculations

Case B3 considers emissions from manhole cover vents over a flowing, partially filled sewer. Air resulting from wind blowing in one end of the sewer is flowing in the upper portion of the sewer. The direction of the airflow relative to the water flow is not considered; it is assumed that the air in the sewer is at thermal and chemical equilibrium with the wastewater at the location of the manhole.

The air velocity resulting from the wind pressure is calculated from a Bernoulli equation based on frictional losses through the unfilled section of the pipe:

$$\nu = \left[\nu_w^2\, \rho_a/\rho_s\, (1 + \frac{fL}{D})\right]^{0.5} = 80 \text{ cm/s} \quad ,$$

where:

ν = linear velocity of air through unfilled section of sewer, cm/s

ν_w = velocity of wind, cm/s

ρ_a = density of ambient air, 0.0012 g/cm^3

ρ_s = density of humid air in sewer, 0.00117 g/cm^3

f = friction factor for air, assumed constant at 0.006, dimensionless

L = length of sewer, 4,570 cm

D = equivalent diameter (four times the hydraulic radius) of unfilled section of sewer, 40.4 cm.

The velocity is then used to calculate the pressure drop through the shorter length of sewer between the manhole and the discharge end of the sewer:

$$\Delta P = \frac{4\, f\, L_s\, \nu^2\, \rho_s}{2\, G_c\, D} = 0.0064 \quad ,$$

where:

ΔP = pressure drop through sewer between manhole and discharge end, g force/cm^2

L_s = length of sewer between manhole and gas exit, 3,050 cm

G_c = gravitational constant, 981 g cm/g force-s^2.

This pressure (0.0064 g force/cm^2) is then used as the driving force in the equation for flow through a square-edged orifice to calculate the linear velocity of air emitted from the manhole cover vents:

$$\nu_c = 0.61\ (2\ G_c\ \Delta P/\rho_s)^{0.5} = 63\ \text{cm/s} \quad ,$$

where:

ν_c = linear velocity through the cover vents, cm/s

0.61 = orifice coefficient (dimensionless) appropriate for the velocity range expected.

Note that the above equations can be combined:

$$\nu_c = 0.61\ (fL_s\ \nu^2/D)^{0.5} = 63\ \text{cm/s} \quad .$$

The linear velocity can be converted to a molar flow rate by multiplying by the cross-sectional area of the vents, 20 cm^2 (four vents each 2.5 cm [1 in.] in diameter assumed in the example), and the molar density of warm humid air at the wastewater temperature, $4 \cdot 10^{-5}$ mol/cm^3. The wastewater flow rate, 2,360 mol/s, has been implicitly specified in the example from the depth, diameter, slope, and roughness of the sewer. The fraction of organics emitted is calculated from the molar flow rates and a dimensionless partition coefficient appropriate for the compound of interest at the wastewater temperature:

$$F = \frac{GK}{GK + L} = 0.008 \quad ,$$

where:

F = fraction of organics emitted through manhole cover vents

G = airflow rate through manhole cover vents, 0.051 mol/s

K = partition coefficient, 371, dimensionless

L = wastewater flow rate, 2,360 mol/s.

B.3.8 Description of Case C1 Calculations

Case C1 considers air blowing directly into one end of a sewer, reaching thermal and compositional equilibrium within the sewer and exiting a junction box.

The maximum pressure exerted by the wind is calculated based on a solution of the Bernoulli equation:

$$dP = \nu^2 \rho/(2 g_c) \quad ,$$

where:

dP = calculated pressure, g force/cm^2
ν = wind velocity, 156 cm/s (3.5 mph)
ρ = density of air at 25 °C, 0.0012 g/cm^3
g_c = 980.665 g-cm/gF-s^2.

$$dP = \frac{(156)^2 \ (0.0012)}{2 \ (980.665)} = 0.015 \text{ g force/cm}^2 \quad .$$

This value of the maximum pressure is equated to the energy of the air velocity in the sewer and the frictional losses in the sewer:

$$dP/\rho = (1 + 4 \ F \ L/D) \ \nu^2/2g_c \quad ,$$

where:

dP = pressure, gforce/cm^2
ρ = density of air, g/cm^3
F = friction factor of air, 0.006
L = length of sewer, 4,570 cm (150 ft)
D = equivalent diameter of the headspace in the sewer, 40.4 cm (four times the hydraulic radius)
g_c = 980.665 g-cm/gF-s^2.

Solving for V, the velocity of air in the sewer is 80 cm/s (1.8 mph). The sectional area of the headspace is 1,828 cm^2, permitting a calculated airflow of 146,000 cm^3/s. The molar density of the air is $4 \cdot 10^{-5}$ mol/cm^3; a molar airflow rate is then calculated as ($1.46 \cdot 10^5$ cm^3/s) ($4 \cdot 10^{-5}$ mol/cm^3), or 5.8 mol/s. The concentration of volatile organics in the air at equilibrium with the initial concentration of volatile organics in the water is as follows:

$$(5.8 \text{ mol/s})(371)(0.0005 \text{ g/g})(18 \text{ g/mol}) \text{ or } 19.4 \text{ g/s} \quad .$$

The fraction of volatile organics present in the air at equilibrium is independent of concentration (as long as K is a constant). The fraction of volatile organics in the air is the ratio of the mass flow in the air divided by the sum of the mass flow in the air and water:

$$f = 19.4/(21.1 + 19.4) = 0.48 \quad .$$

General assumptions and calculations:

Air temperature	25 °C
Relative humidity	50 percent
Sewer temperature	30 °C
Friction factor for air	0.006
Wind velocity	156 cm/s (3.5 mph)
Radius of sewer	30.48 cm (12 in.)
Depth of liquid in sewer	24.4 cm (9.6 in.)
Headspace hydraulic radius	10.9 cm
Flow of water in sewer	42,196 cm^3/s
Headspace area in sewer	1,828 cm^2
Density of air at 25 °C	0.0012 g/cm^3
K partition coefficient (Y/X)	371
Weight fraction volatile organics in water	0.0005
Flow of volatile organics in sewer water	21.1 g/s
Molar density of air in sewer	0.00004 mol/cm^3.

B.3.9 Description of Case C2 Calculations

Case C2 estimates airflow into a sewer from a junction box, induced by water flow in the sewer. This air reaches thermal and compositional equilibrium within the sewer and is discharged from the sewer at the next junction box.

The velocity profile for the surface of the water in the sewer is assumed to be given by the following empirical relationship:

$$\nu^{+} = 8.5 + 2.5 \ln(Y/e) \quad , \qquad [Y > e]$$

where:

ν^{+} = velocity quotient, the ratio of the velocity to the friction velocity

e = surface roughness, cm

Y = distance from a point on the surface to the nearest wall-surface interface, cm.

The average velocity in the sewer is estimated as 38.7 cm/s integrating the above equation for average flow. The average surface velocity was 42.2

cm/s. The perimeter of the surface was 60 cm, and the perimeter of the sewer headspace was 108 cm. The average velocity of the airflow was established as follows:

$$(\text{cm/s}) = 42.2 \text{ cm/s} \left(\frac{60 \text{ cm}}{60 \text{ cm} + 108 \text{ cm}}\right) \quad .$$

This average air velocity is 36 percent of the water velocity in the sewer. This estimated ratio of air velocity to water velocity compares favorably to a reported percentage of 35 by Pescod and Price (Journal WPCF, vol. 54, no. 4 (April 1982), p. 393) for laminar airflow due to liquid drag. The estimated Reynolds number for the above flow conditions suggests that the flow of air may be in the transitional zone. The assumption of laminar flow of air may have overestimated the flow of air by 20 percent.

The estimated velocity of air in the sewer is 15 cm/s (0.33 mph). The sectional area of the headspace is 1,828 cm^2, permitting a calculated airflow of 27,000 cm^3/s. The molar density of the air is $4 \cdot 10^{-5}$ mol/cm^3; a molar airflow rate is then calculated as ($2.7 \cdot 10^4$ cm^3/s)($4 \cdot 10^{-5}$ mol/cm^3), or 1.08 mol/s. The flow rate of volatile organics in the air at equilibrium with the initial concentration of volatile organics in the water is as follows:

$$(1.08 \text{ mol/s})(371)(0.0005 \text{ g/g})(18 \text{ g/mol}) \text{ or } 3.61 \text{ g/s} \quad .$$

The fraction of volatile organics present in the air at equilibrium is independent of concentration (as long as K is a constant). The fraction of volatile organics in the air is the ratio of the mass flow in the air divided by the sum of the mass flow in the air and water:

$$f = 3.61/(21.1 + 3.61) = 0.146 \quad .$$

General assumptions and calculations:

Air temperature	25 °C
Relative humidity	50 percent
Sewer temperature	30 °C
Friction factor for air	0.006

Radius of sewer	30.48 cm (12 in.)
Depth of liquid in sewer	24.4 cm (9.6) in.)
Headspace hydraulic radius	10.9 cm
Flow of water in sewer	42,196 cm^3/s
Headspace area in sewer	1,828 cm^2
Density of air at 25 °C	0.0012 g/cm^3
K partition coefficient (Y/X)	371
Weight fraction volatile organics in water	0.0005
Flow of volatile organics in sewer water	21.1 g/s
Molar density of air in sewer	0.00004 mol/cm^3
Reynolds number for airflow	2,110
Average velocity of water	39 cm/s
Surface velocity of water	42 cm/s
Roughness of sewer wall	0.21 cm (0.007 ft)
Slope of sewer	0.000431.

B.3.10 Description of Case C3 Calculations

This calculation considers airflow from the discharge end of a partially filled sewer to the influent end of the sewer resulting from a density difference between the cooler ambient air and the warm humid air in equilibrium with the wastewaters. The air flowing from the sewer is assumed to be in thermal and chemical equilibrium with the wastewater. Air and water flow countercurrently.

The ambient temperature and relative humidity and the wastewater temperature are used to calculate the density difference; the slope and length of the sewer are used to calculate the elevation difference producing the "stack effect." Based on the length, diameter, and depth in the sewer, the frictional resistance to airflow is determined as a function of air velocity. The air velocity is calculated from a balance of the "stack effect" and the frictional losses using a form of the Bernoulli equation:

$$\nu = [2\ g_c\ \Delta\rho h/\rho/(1 + 4\ FL/D)]^{0.5} \quad ,$$

where:

ν = velocity of air through the sewer headspace, 5.2 cm/s

g_c = acceleration of gravity, 981 cm/s^2

$\Delta\rho$ = density difference between ambient air and warm humid air in sewer, $3.2 \cdot 10^{-5}$ g $force/cm^2$

h = elevation difference determined from sewer length and slope, 6.6 cm

ρ = density of warm humid air in sewer, 0.00117 g/cm^3

F = friction factor for airflow through sewer, assumed constant at 0.006, dimensionless

L = sewer length, 4,570 cm

D = diameter of circle having equivalent area to the cross section of the sewer headspace, 40.4 cm.

This velocity is converted to a molar flow rate by multiplying by the cross-sectional area of the headspace in the sewer, 1,828 cm^2, and the molar density of air at the wastewater temperature, 4.0•10^{-5} mol/cm^2.

Water flow rate (2,360 mol/s) is specified implicitly by the slope, diameter, depth, and roughness of the sewer. The fraction of influent volatile organics that is emitted is calculated from the molar flow rates of water and air and the dimensionless partition coefficient for the compound of interest at the wastewater temperature. Air emitted from the sewer is assumed to be in equilibrium with influent wastewater. The fractional emissions are calculated as:

$$F = \frac{GK}{GK + L} = 0.057 \quad ,$$

where:

G = airflow rate, 0.382 mol/s

K = 371, dimensionless partition coefficient for compound of interest

L = wastewater flow rate, 2,360 mol/s.

B.3.11 Description of Case D1 Calculations

Case D1 considers emissions from a stack on a sump. The stack was designed to promote the discharge of fumes above workers' heads so that their exposures to environmental releases would be reduced. Case D1 uses a method identical to Case A3 to estimate the airflow due to the stack effect.

Case D1 considers airflow up from a sump through a vent induced by density differences between the ambient air outside the sump and the warm humid air in the sewer. The wastewater in the sewer is assumed to be flowing in a direction perpendicular to the airflow through the vents; the air is assumed to be saturated with water and at chemical and thermal equilibrium with the wastewater. In the case considered, the vent is assumed to be 10 cm (4 in.) in diameter and 366 cm (12 ft) long. Frictional losses through both the drain and the sewer are considered, based on a friction factor of 0.06. The height of the "stack" is assumed to be 366 cm (12 ft). This is the vertical distance between the top of the sump and the vent top. Ambient conditions are assumed to be 25 °C and 50 percent relative humidity. The wastewater temperature is assumed to be 30 °C.

The densities of ambient air and warm humid sewer air are calculated, and the density difference across the vent system is calculated as $3.2 \cdot 10^{-5}$ g/cm^3. The maximum pressure from density differences is the product of the density difference and height. This value of the maximum pressure from density differences is equated to the energy of the air velocity in the sewer and the frictional losses in the sewer:

$$dP/\rho = (1 + Ke + 4\ F\ L/D\ Arr^2 + 4\ F\ L2/D2 + K1)\ \nu^2/2g_c \quad ,$$

where:

- dP = pressure, 0.0117 g force/cm^2
- ρ = density of air, 0.0012 g/cm^3
- Ke = diameter change coefficient, 0.378
- F = friction factor of air, 0.006
- L = length of sewer, 3,048 cm
- D = equivalent diameter of the headspace in the sewer, 43.6 cm (four times the hydraulic radius)
- Arr = area ratio of sewer segment, 0.055
- $L2$ = length of drain, 366 cm
- $D2$ = diameter of stack, 10 cm
- $K1$ = loss coefficient, 3
- g_c = 980.665 g-cm/gF-s^2.

Solving for V, the velocity of air in the sewer is 11 cm/s (21 ft/min). The sectional area of the headspace is 1,830 cm^2, permitting a calculated airflow of 20,000 cm^3/s. The molar density of the air is $4.0 \cdot 10^{-5}$ mol/cm^3; a molar airflow rate is then calculated as (20,000 cm^3/s)($4.0 \cdot 10^{-5}$ mol/cm^3), or 0.8 mol/s. The flow rate of volatile organics in the air at equilibrium with the initial concentration of volatile organics in the water is as follows:

$$(0.8 \text{ mol/s})(371)(0.0005 \text{ g/g})(18 \text{ g/mol}) \text{ or } 2.7 \text{ g/s} \quad .$$

The fraction of volatile organics present in the air at equilibrium is independent of concentration (as long as K is a constant). The fraction of volatile organics in the air is the ratio of the mass flow in the air divided by the sum of the mass flow in the air and water:

$$f = 2.7/(21.3 + 2.7) = 0.11 \quad .$$

General assumptions and calculations:

Air temperature	25 °C
Relative humidity	50 percent
Sewer temperature	30 °C
Friction factor for air	0.006
Radius of sewer	30.48 cm (12 in.)
Depth of liquid in sewer	24.4 cm (9.6 in.)
Headspace hydraulic radius	10.9 cm
Flow of water in sewer	42,000 cm^3/s
Headspace area in sewer	1,830 cm^2
Density of saturated air at 40 °C	0.00117 g/cm^3
Density of air at 25 °C	0.0012 g/cm^3
K partition coefficient (Y/X)	371
Weight fraction volatile organics in water	0.0005
Flow of volatile organics in sewer water	21.3 g/s
Molar density of air in sewer	0.00004 mol/cm^3.

B.3.12 Description of Case D2 Calculations

Case D2 considers emissions from open trenches around process equipment. These trenches are used to collect process wastes, tank cleaning wastes, unplanned leaks, and water. Air blows across the top of a grate covering the open top channel of the trench.

The depth of flow in the channel is 7.6 cm (3 in.), the velocity is 76 cm/s (1.5 ft/s), and the length is 12.2 m (40 ft). The mass transfer coefficient of the gas phase is assumed to be one-half the value for surface impoundments (Kg = 0.0065/2 m/s). The more rapid flow would tend to increase the mass transfer coefficient and the grate covering would tend to decrease the mass transfer coefficient. The liquid mass transfer coefficient is calculated by Owens, Edwards, and Gibbs (<u>Int. J. Air Wat. Poll.</u>, vol. 8 (1964), pp. 469-486):

$$Kal = (21.6\ V^{0.67}/H^{1.85})(K/Ko) \quad ,$$

where:

Kal = liquid mass transfer coefficient, 176 /day,
V = velocity, 1.5 ft/s,
H = depth, 0.25 ft, and
K/Ko = ratio of mass transfer coefficients of toluene and air, 0.477.

The overall mass transfer coefficient is obtained by summing the resistance of the two regions of mass transfer in series:

$$1/Ka = 1/Kal + H/(C\ Kg\ K \bullet 0.000736) \quad ,$$

where:

Ka = 149/day
Kal = 280
H = 0.25 ft
C = 24•3,600/12/2.54•100 (ft/day) (s/m)
Kg = 0.00325 m/s
K = 371 (Y/X).

The residence time in a 40-ft length of channel is 40 ft (1.5 ft/s) or 27 s or $3.1 \bullet 10^{-4}$ days. The fraction lost during flow through the 40-ft channel is estimated with the following equation:

$$f = 1 - EXP\ (-Ka\ t) = 0.045 \quad .$$

Therefore, 4.6 percent of toluene is estimated to be emitted over the 40-ft section of channel.

B.4 A COMPARISON OF THEORETICAL PREDICTIONS TO MEASURED VALUES

Several industrial plants were visited that have wastewater collection systems. Screening measurements were made at these sites with an Alnor velometer (low-velocity probe) to evaluate the magnitude of sewer emissions. Of primary concern was to determine the magnitude of the measured airflow velocities and to compare these measured velocities with the predicted velocities.

Based upon the results of the velocity screening measurements, the following observations can be stated:

- The velocities from the openings in the wastewater collection systems were variable, but the general range of velocities from the different sources overlapped (Table B-8).
- The preliminary results of using the model for site specific conditions was favorable, with reasonable agreement (factor of 2) between the predicted and measured velocities (Table B-9).
- To improve the agreement between the model results and the observed sewer velocities, representative model plant parameters were used in the theoretical estimations. Examples of sources that were modified include open drains under grates (higher emissions than predicted) and sealed drain systems (lower emissions than predicted).
- Some of the sealed sewer systems that were observed could be considered as a control technology relative to other more open sewer systems.

General comments offered about the plant visits include the following:

- The collection system at the first refinery may not have been representative of collection systems at other chemical plants and refineries. The refinery representative said that a substantial amount of effort had been directed to reducing collection system emissions.
- The collection systems at the paper mills may be more typical of chemical plants than the systems at the first refinery. Open-grated drains may be common around process equipment, perhaps because they can trap wastes from both unexpected spills and planned cleanups.

TABLE B-8. SCREENING VALUES FOR AIR VELOCITIES AT SEWER OPENINGS

Unit	Location	Velocity at opening[a] ft/min	ft/s
Chemical sewer	Refinery 1		3
Drain grate	Refinery 1		2
Chemical sewer	Refinery 1	80	1.3
Open drain	Refinery 1	65	1.1
Manhole (1-in. dia. opening)	Refinery 1	200	3.3
Open drain	Refinery 1	40	0.66
Closed drain	Refinery 1	60	1.0
Opening	Refinery 1	120	2
Sump opening	Refinery 1		0-2
Sample point drain	Refinery 1	150	2.5
Sample point drain	Refinery 1	0[b]	
Chemical sewer sump	Refinery 1	110	1.8
Drain opening	Refinery 1		0-1
Horizontal flow in sewer	Pulp Mill 2	50	0.83
Manhole cover	Pulp Mill 2	300	5.0
Manhole cover	Pulp Mill 2	100-150	1.7-2.5
Grate over sewer	Pulp Mill 3	50-100	0.83-1.7
Grate	Pulp Mill 3	50	0.83
Lift station opening	Pulp Mill 3	40-50	0.67-0.83
Lift station opening	Pulp Mill 3	60	1
Floor drain (2 x 1 ft)	Pulp Mill 4	50	0.83
Floor trench (1-2 ft/s)	Pulp Mill 4	50	0.83
Floor trench	Pulp Mill 4	0-50	0-1
Main drains between process units	Pulp Mill 4	100	1.7
Small vent on main sewer	Pulp Mill 4	>300	5
Grate on main sewer	Pulp Mill 4	150	2.5
Grate on main sewer	Pulp Mill 4	160-170	2.7-2.9
Grate open drain	Pulp Mill 4	50	0.83
Grate at end of trench	Pulp Mill 5	100	1.7

[a]Measured with an Alnor velometer, low-velocity probe. Feet per second were reported without corresponding feet per minute obtained by visual inspection of plume rise.

[b]No measured velocity, but visual observation of slow fumes leaving drain.

TABLE B-9. A COMPARISON OF MEASURED AND PREDICTED AIR VELOCITIES AT SEWER OPENINGS

Unit	Location	Velocity at opening (ft/min)[a]	
		Measured	Predicted
Drain	Refinery 1	65,40,60,120	122 (A2)[b]
Manhole cover	Pulp Mill 2	300	124 (B3)
Manhole cover	Pulp Mill 2	100-150	191 (B2)
Average drain velocities		67	84
Average manhole cover opening velocities		198	128
Average junction opening velocities		88	66

[a]Measured with an Alnor velometer, low-velocity probe. Predicted values based on a wind velocity of 3.5 mph (300 ft/min).

[b]Wind effect (3.5 mph) probably overestimates emissions because actual wind velocity was lower (tall process units, wind direction different from stack emissions at high elevation).

Appendix C—Physical/Chemical Properties

TABLE C-1. TARGETED ORGANIC COMPOUNDS AND THEIR PHYSICAL PROPERTIES AT 25°C[a]

Organic Compound	Molecular Weight (g/gmol)	Density (g/cm^3)	Henry's Constant (atm-m^3/gmol)	D_w (cm^2/sec[a])	D_a (cm^2/sec)
Toluene	92.0	0.87	6.68×10^{-3}	8.6×10^{-6}	8.7×10^{-2}
Naphthalene	128.2	1.14	1.18×10^{-3}	7.5×10^{-6}	5.9×10^{-2}
Phenol	94.1	1.07	4.54×10^{-7}	9.1×10^{-6}	8.2×10^{-2}
1,3-butadiene	54.1	0.62	1.42×10^{-1}	1.08×10^{-5}	2.49×10^{-1}
1-Butanol	74.1	0.81	8.90×10^{-6}	9.30×10^{-5}	8.00×10^{-2}

[a]Reference 1.

TABLE C-2. ORGANIC COMPOUND PROPERTIES

Compound	Vapor Pressure 25°C (mm Hg)	Vapor Pressure 100°C (atm)	Partition[a] Coefficient 25°C (Y/X)	Partition[a] Coefficient 100°C (Y/X)	Biorate[d] (mg/g biomass-hr)
1,3-Butadiene	2095	16.7	7889[b]	47717[c]	0.39
Toluene	30	0.73	371	279	73.48
Naphthalene	0.23	0.024	66	28.7	42.47
1-Butanol	6.5	0.51	0.49	9.4	32.43
Phenol	0.341	0.054	0.025	7.7	97.0

[a]Reference 2.

[b]Estimated based on Henry's Law constant (H):

Partition coefficient (K) = H (atm·m^3/gmol) (1 x 10^6 cm^3/m^3)

(1/1 atm total pressure) (1 gm H_2O/cm^3)(1 gmol H_2O/18 gm H_2O)

[c]Estimated from: K_{25} VP_{100}/VP_{25}

where K_{25} = partition coefficient at 25°
VP_{25} = vapor pressure at 25°C
VP_{100} = vapor pressure at 100°C

[d]Reference 1.

TABLE C-3. PARTITION COEFFICIENTS FOR SELECTED ORGANIC COMPOUNDS AT VARIOUS TEMPERATURES[a]

			Ki Values				
		Temperature °C	1,3-Butadiene	1-Butanol	Naphthalene	Phenol	Toluene
I.	Purge and Trap						
	8010/846	25	7,889	0.49	65.56	0.025	371.11
	8030/846	85	3,602	7.62	36.1	6.17	297
	EPA/RTI	25	7,889	0.49	65.56	0.025	371.11
	EPA 624	25	7,889	0.49	65.56	0.025	371.11
	CARB 401		23.5	0.239	0.0345	0.00486	0.639
II.	Steam Distillation						
	EPA	100	12,672	9.4	28.7	7.7	279
	AQMD	100	12,672	9.4	28.7	7.7	279
III.	Headspace						
	EPA/TRC	25	7,889	0.49	65.56	0.025	371.11
	EPA/TRC	90	39,370	8.21	33.6	6.68	291

[a]Reference 2.

C-1. REFERENCES

1. Office of Air Quality, Planning and Standards. U.S. Environmental Protection Agency. Hazardous Waste Treatment, Storage, and Disposal Facilities (TSDF) - Air Emission Models, April 1989.

2. "Preliminary Evaluation of Test Methods for Volatile Organics in Hazardous Wastes: Batch Steam Stripping/Distillation." Office of Air Quality Planning and Standards, U.S. Environmental Protection Agency, February 4, 1988.

Appendix D—Steam Stripper Performance Data

TEST DATA SUMMARY — SITE A

1.0 FACILITY DESCRIPTION

Wastewater from the production of 1,2-dichloroethane is collected in a feed tank, from which the waste is pumped into the steam stripper column. The organics are stripped from the waste and condensed overhead in a series of two condensers. The entire condensate, both aqueous and organic phases, is recycled to the production process. The effluent stream from the stripper column is sent through a heat exchanger to help preheat the feed stream and is then sent to a wastewater treatment facility.

A schematic of the steam stripping system is shown in Figure 1. No design information is available for the tray steam stripper column. Typically, the feed rate is about 850 L/min to the column operating at 136 kiloPascals (kPa). Steam is fed at 446 kPa and unspecified temperature at a rate of about 1,700 kg/hr.

2.0 STEAM STRIPPER SAMPLING AND ANALYSIS

Sampling and process measurement locations are detailed on the process diagram in Figure 1. Sampling was conducted over two days with samples taken five times at 2-hour intervals on each day. Liquid grab samples were collected in 40-ml volatile organic analyzers (VOA) vials. Gas vent samples were collected in evacuated stainless steel canisters. Process data were collected during the testing at half-hour intervals throughout the testing. Process operation data collected included feed, effluent, condensate, and steam flow rates, temperatures of the feed, effluent, and condensate, and the steam pressure.

The organic compounds in the water samples were analyzed by a purge and trap procedure with separation and quantification performed by gas chromatography/mass spectroscopy (GC/MS) analysis (EPA Method 624). The organic phase in the condensate was analyzed by direct injection gas chromatography. The vent gas analysis procedures were detailed in the site-specific test and quality assurance plan dated July 7, 1986, but were not presented in the report.

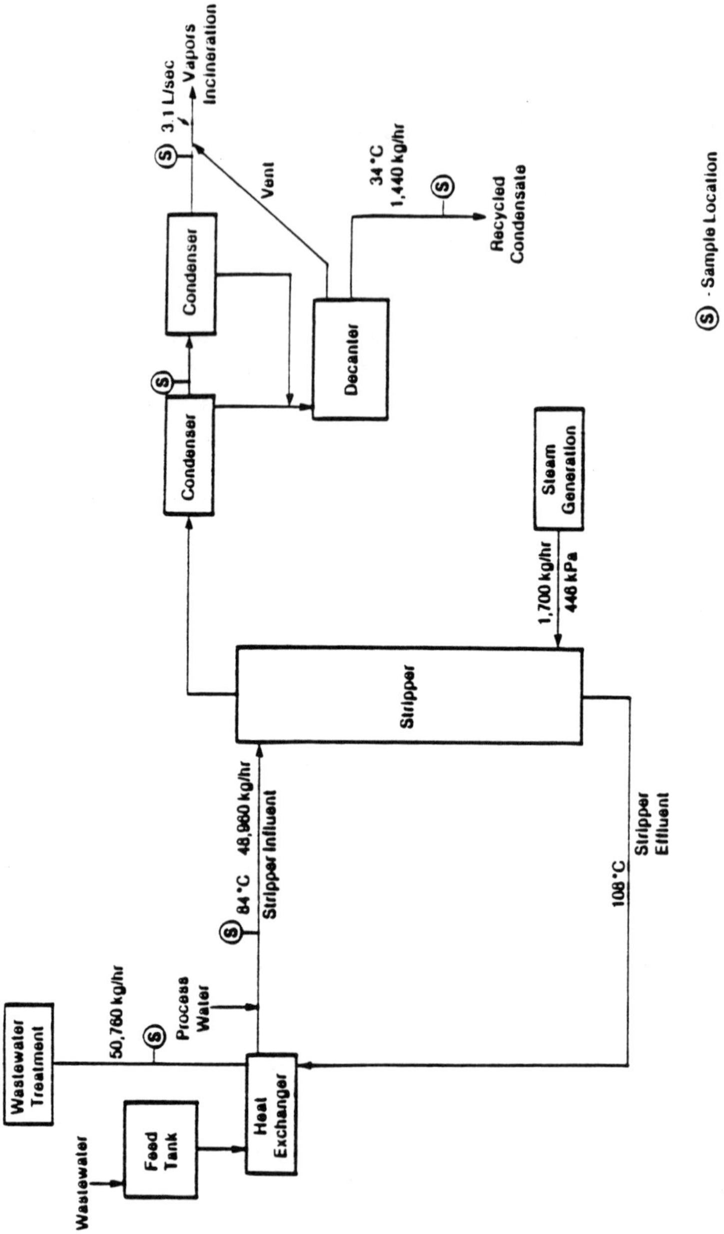

Figure 1. Diagram of Plant A steam stripper system and sampling locations.

Stream flow and concentration data were used to characterize all process streams around the steam stripper. Averages of all flow and concentration data taken were used in preparing the summaries of these data as shown in Tables 1 and 2. Table 1 presents the flow rate, temperature, and pressure of streams entering and leaving the steam stripper. Table 2 presents composition data for each of these streams as well as organic removal efficiencies. The overhead removal percentage was calculated on the basis of influent and effluent flows from the stripper. The composition data available for the condensate is presented in Table 2, but is not used to calculate removal efficiencies. This is done because of the need to see the actual amount of organic removed from the wastewater and because of the incompleteness of the condensate data.

Site A

TABLE 1. BULK STREAM CHARACTERIZATION

Parameter	Influent to Stripper[a]	Effluent from Stripper[b]	Overhead Condensate[c]		Steam	Influent Steam
			Aqueous	Organic		
Flow Rate (kg/hr)[d]	48,960	50,760	1,200	240	1,700	-
Temperature (°C)	84	108	34	34	NA	-
Pressure (kPa)	NA	136	NA	NA	446	-
Unit Ratio (kg/kg)	-	-	-	-	-	28.8

NA - Not Available

[a]Shown as Stripper Influent in Figure 1.
[b]Shown as Stripper Effluent in Figure 1.
[c]Shown as Recycled condensate in Figure 1. The entire condensate stream is recycled, but the organic and aqueous flows are presented separately here.
[d]Calculated as average of all flow rates measured during testing.

Site A

TABLE 2. COMPONENT STREAM CHARACTERIZATION

Component	Influent to Stripper Flow (kg/hr)	Influent to Stripper Conc.[a] (ppmw)	Effluent from Stripper Flow (kg/hr)	Effluent from Stripper Conc.[a] (ppmw)	Overhead Condensate[b] Aqueous Flow (kg/hr)	Overhead Condensate[b] Aqueous Conc.[a] (ppmw)	Overhead Condensate[b] Organic Flow (kg/hr)	Overhead Condensate[b] Organic Conc.[a] (ppmw)	Overhead Removal[c] (Wt.%)
1,2-Dichloroethane	270[d]	5,600	0.0049	0.097	4.3	3,600[e]	210	880,000[e]	99.998
Chloroform	13	270	0.48	9.6	0.41	340[e]	11	47,000[e]	96
Benzene	0.0098	0.20	2.5E-4	0.005[f]	-	NA	-	NA	97
Carbon Tetrachloride	0.083	1.7	2.5E-4	0.005[f]	-	NA	-	NA	99.7
Chlorobenzene	0.017	0.34	2.5E-4	0.005[f]	-	NA	-	NA	99
Chloroethane	0.47	9.6	2.5E-4	0.005[f]	-	NA	-	NA	99.95
1,1-Dichloroethane	0.54	11	2.5E-4	0.005[f]	-	NA	-	NA	99.95
1,1-Dichloroethene	0.23	4.7	2.5E-4	0.005[f]	-	NA	-	NA	99.9
1,2-dichloroethene	0.44	8.9	2.5E-4	0.005[f]	-	NA	-	NA	99.9
Methylene Chloride	0.059	1.2	2.5E-4	0.005[f]	-	NA	-	NA	99.6
Tetrachloroethene	0.069	1.4	2.5E-4	0.005[f]	-	NA	-	NA	99.6
1,1,2-Trichloroethane	0.37	7.5	2.5E-4	0.005[f]	-	NA	-	NA	99.9
Trichloroethene	0.24	4.8	2.5E-4	0.005[f]	-	NA	-	NA	99.9
Vinyl Chloride	0.41	8.4	2.5E-4	0.005[f]	-	NA	-	NA	99.9
Total VO	290	5,900	0.50	9.8	4.7	3,900	220	927,000	99.8
Water	48,670[g]	-	50,760[g]	-	1,200[g]	-	20[g]	-	-
Total	48,960[h]	-	50,760[h]	-	1,200[h]	-	240[h]	-	-

NA - Not analyzed for this component

[a]Average of concentrations measured during testing.

[b]Not used for calculation of removal efficiencies because of need to determine actual organic removed and incompleteness of condensate analyses.

[c]Calculated as: (1 - (Effluent from stripper)/(Influent to stripper)) x 100%.

[d]Sample calculation: (Average component concentration, ppmw) x (Average flow rate) x $(10^6 \text{ ppmw})^{-1}$ = Component flow rate.

[e]Only chloroform and 1,2-dichloroethane were analyzed in the condensate. Because of the use of average flows and average concentrations, the component mass balance for these components may not be close as was usually obtained at a given sampling time.

[f]All concentrations were below detection limit. One-half of limit used for calculation purposes.

[g]Balance of flow after accounting for organics.

[h]Average of flow rates measured during testing.

TEST DATA SUMMARY— SITE B

1.0 FACILITY DESCRIPTION

A process waste stream consisting of methylene chloride, water, salt, and organic residue is fed to the steam stripper in which much of the organic compounds is stripped and taken overhead. The overhead vapor is condensed, with the aqueous phase being recycled to the column and the organic phase stored for reuse. The bottom stream is used to preheat the incoming waste and is then either sent to a publicly owned treatment works or back into a tank for the feed stream, depending on whether the effluent meets discharge limits. If the midpoint temperature of the stripping column is above a given set point, the effluent meets limitations and is sent to the treatment facility.

A schematic of the steam striper system is shown in Figure 1. The stripping column contains 10 feet of 5/8 inch pall rings and has a diameter of eight inches. The waste stream feed rate is approximately 19 L/min with an overhead organic product rate of about 0.28 L/min. Steam was fed at a pressure range of 190-320 kiloPascals (kPa), although the temperature and rate were unspecified.

2.0 STEAM STRIPPER SAMPLING AND ANALYSIS

Desired sampling and process measurement locations are detailed on the process diagram in Figure 1. All process data recorded during sampling are presented also. Sampling of the influent and effluent was conducted approximately hourly for five hours on the first day and for 12 hours on the second day, although a shut down and restart delay of six hours occurred the second day because of instrument difficulties. Liquid grab samples were collected in either a glass or stainless steel beaker and then distributed into individual glass bottles for analysis. A composite sample of the organic product was collected in glass bottles after completion of the test. Gas vent samples were collected in evacuated glass sampling bulbs. Process data collected included feed flow rate, column, feed, effluent and vent temperatures, and steam pressure.

Vent gas was analyzed using chromatography with a flame ionization detector (GC/FID) (Method 18). The volatile organic compounds in the liquid samples were analyzed by gas chromatography/mass spectrometry (GC/MS)

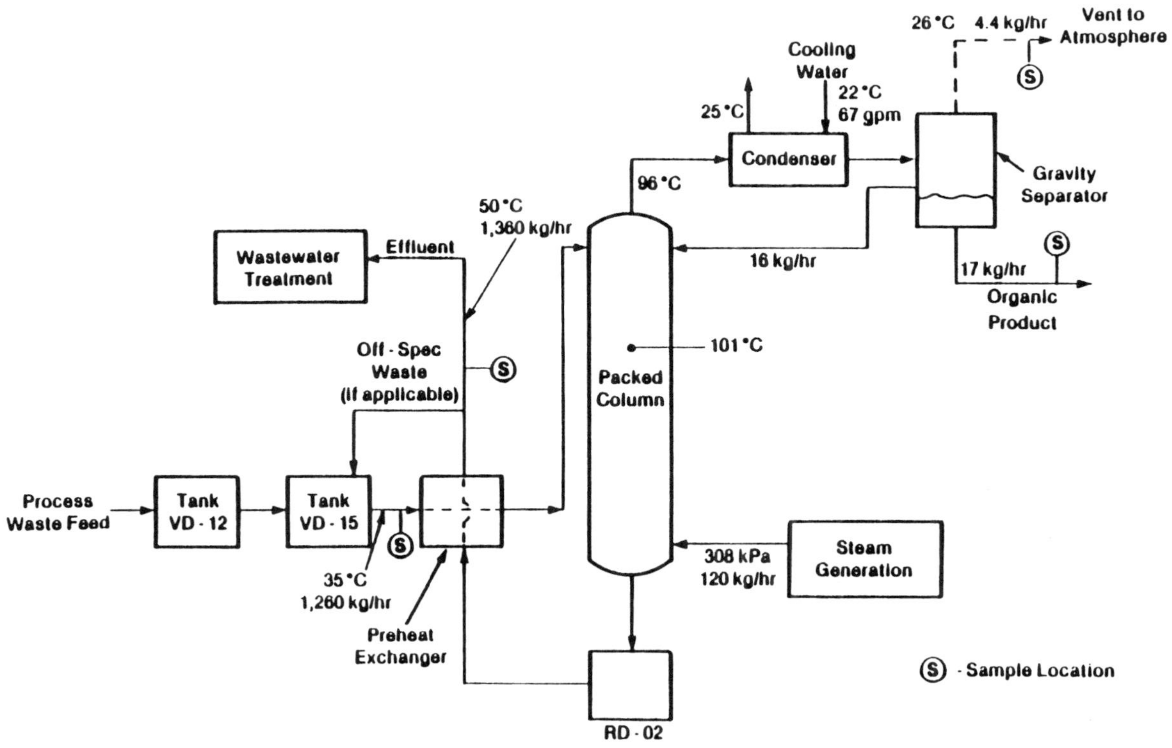

Figure 1. Diagram of Steam Stripper at Site B with Sampling Locations.

(Method 8240).` Material and energy balances and stream flow and concentration data were used to characterize all process streams around the steam stripper. Summaries of the flow and concentration data gathered during the test are shown in Tables 1 and 2. Table 1 presents the flow rate, temperature, and pressure of streams entering and leaving this steam stripper process. Table 2 presents composition data for each of these streams as well as organic removal efficiencies.

SITE B

TABLE 1. BULK STREAM CHARACTERIZATION

Parameter	Influent[a]	Effluent[b]	Organic Overhead Condensate[c]	Steam	Influent/ Steam
Flow Rate (kg/hr)	1,260	1,360[d]	17[f]	120[d]	-
Temperature (^oC)	35	50	NA	NA	-
Pressure (kPa)	NA	NA	NA	308[e]	-
Unit Ratio (kg/kg)	-	-	-	-	10.5

NA - Not Available

[a]Shown as Process Waste Feed in Figure 1.
[b]Shown as Effluent in Figure 1.
[c]Shown as Organic Product in Figure 1.
[d]Calculated from material and energy balances using measured flows.
[e]Steam used at range of 190 to 320 kPa. This value selected as weighted average.
[f]Volume recovered estimated in report. Flow rate calculated based on measured specific gravity and elapsed times for given volume.

Site B

TABLE 2. COMPONENT STREAM CHARACTERIZATION

Component	Influent Flow (kg/hr)	Influent Concentration (ppmw)	Effluent Flow (kg/hr)	Effluent Concentration (ppmw)	Overhead Condensate[f] Organic Flow (kg/hr)	Overhead Condensate[f] Organic Concentration (ppmw)	Overhead Removal[g] (Wt.%)
Methylene Chloride	4.5[c]	3,600[d]	0.0003[e]	0.22	0.088	5,200	99.99
Chloroform	0.066	52	0.0072	5.3	0.019	1,100	89
Carbon Tetrachloride	0.0019	1.5	0.0001[e]	0.09	2.1E-5[e]	1.2	95
Total VO	4.6	3,600	0.0076	5.6	0.107	6,300	98
Water	1,255[a]	-	1,360[a]	-	16.9[a]	-	-
Total	1,260	-	1,360[b]	-	17	-	-

[a]Balance of total flow after accounting for organics.
[b]Calculated from material and energy balances using measured flows.
[c]Sample calculation:
(Component Concentration, ug/L) x (Flow rate, L/hr) x (10^{-9} kg/ug) = Component flow rate.
Component Concentration is an average of the grab samples taken during testing.
[d]Calculated from component and total flow rates.
[e]Some concentrations observed were below the detection limit. One-half the detection limit used for calculating the average concentration.
[f]Not used for calculation of removal efficiencies because of desire to see actual removal from waste stream and to remove any background interference effects.
[g]Calculated as:
(1 - (Effluent flow)/(Influent flow)) x 100%.

TEST DATA SUMMARY — PLANT G

1.0 FACILITY DESCRIPTION

Wastewater from a feed tank is pumped to the steam stripping column where the organics are steam stripped in the column and condensed in the overheads. The stripped organics are separated from the condensed steam in the organic condensate tank. The aqueous layer is recycled from the organic condensate tank to the feed tank. The organic phase is sent to a vented storage tank. From there, the organics are transferred to tank trucks and taken off site for resale as fuel. Effluent from the steam stripper is passed through a liquid-phase carbon adsorption unit to recover any residual organics in the stream. The effluent is then pH adjusted and discharged to surface water.

A schematic of the steam stripping system is shown in Figure 1. The steam stripping column is 19.2 meters high with an internal diameter of 0.46 meters. The column is packed with 3.17 m^3 of 2.5 cm diameter stainless steel rings. The steam stripper operates with a liquid to gas (L/G) ratio ranging from 55 m^3/m^3 at the bottom of the column to 24 m^3/m^3 at the top of the column. Steam is fed to the column at approximately 130°C and 365 kiloPascals (kPa) pressure at a feed to steam ratio of 14.7 kg/kg.

2.0 STEAM STRIPPER SAMPLING AND ANALYSIS

Sampling and process measurement locations are detailed on the process diagram in Figure 1. Sampling was conducted over a 2½ hour period with an average of four samples collected from each sampling point. Liquid grab samples were collected in 40 ml volatile organic analyzer (VOA) bottles. Gas vent samples were collected in evacuated stainless steel canisters. Process operating data were collected over a 4½ hour period to ensure that the process was operating at steady state. Process data collected included feed, steam, and vent gas flow rates, temperatures, and pressures.

Vent gas was analyzed using GC headspace analysis method (Method 5020). The volatile organic compounds in the liquid samples were speciated and quantified by direct injection GC. Confirmation of GC peak identification was carried out by gas chromatography/mass spectrometry (GC/MS) on selected samples (Method 8240). Material and energy balances and stream flow and concentration data were used to characterize all process streams around the

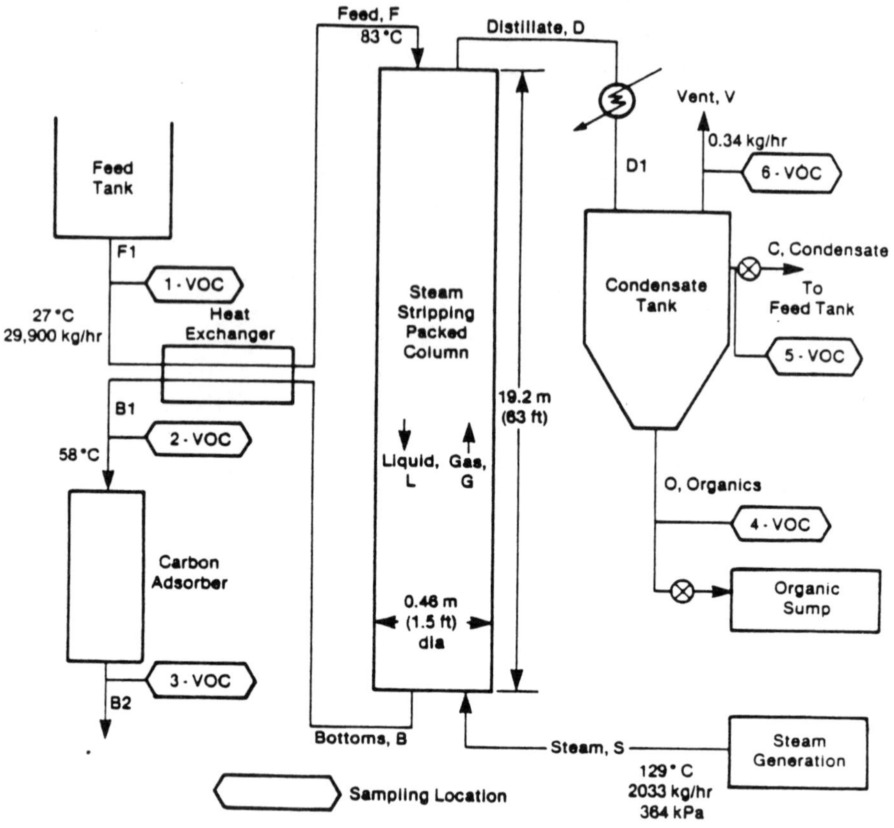

Figure 1. Diagram of Plant G Steam Stripping Process with Sampling Locations

steam stripper and carbon adsorption unit. Summaries of the flow and concentration data gathered during the test are presented in Tables 1 and 2. Table 1 presents the flow rate, temperature and pressure of streams entering and leaving the steam stripper and carbon adsorption unit. Table 2 presents composition data for each of these streams as well as organic removal efficiencies.

The steam stripper removal percentage was calculated based on the influent and effluent flows for the stripper. The composition data for the overhead streams is present, but is not used to calculate removal efficiencies. This is done to show the actual removal of organics from the waste stream. This also minimizes any background interference effects for the wastewater. By looking at the same bulk stream of liquid, the same liquid background is present, allowing for consistency between samples.

PLANT G STEAM STRIPPER

TABLE 1. BULK STREAM CHARACTERIZATION

Parameter	Influent to Stripper[a]	Effluent from Stripper[b]	Effluent from Carbon Adsorber[f]	Overhead Condensate Aqueous[c]	Overhead Condensate Organic[d]	Steam	Influent/Steam
Flow Rate (kg/hr)	29,900	31,490	31,490	427.6	15.4	2,033	-
Temperature (°C)	83	112	58	100[e]	100[e]	128.8	-
Pressure (kPa)	-	204.3	204.3	101.3	101.3	364.3	-
Unit Ratio (kg/kg)	-	-	-	-	-	-	14.7

[a]Shown as Feed, F in Figure 1
[b]Shown as Bottoms, B in Figure 1
[c]Shown as Condensate, C in Figure 1
[d]Shown as Organic, O in Figure 1
[e]Assumed to operate at steam condensation temperature
[f]Shown as B2 in Figure 1

PLANT G STEAM STRIPPER

TABLE 2. COMPONENT STREAM CHARACTERIZATION

Component	Influent		Effluent: From Stripper		Effluent: From Carbon Adsorber		Overhead Condensate[f]: Aqueous		Overhead Condensate[f]: Organic		Steam Stripper Removal[g] (Wt.%)	Carbon Adsorber Removal[e] (Wt.%)
	Flow (kg/hr)	Conc. (ppmw)	Flow (kg/hr)	Conc. (ppmw)	Flow (kg/hr)	Conc. (ppmw)	Flow (kg/hr)	Conc. (ppmw)	Flow (kg/hr)	Conc. (ppmw)		
Nitrobenzene	15.1[d]	505	1.29	41	<0.025	<0.8	0.812	1,900	12.12	787,000	91.5	>98
2-Nitrotoluene	2.33	78	0.076	2.4	<0.025	<0.8	0.037	87	2.97	193,000	96.7	>67
4-Nitrotoluene	1.53	51	0.139	4.4	<0.025	<0.8	0.019	45	1.49	97,000	90.9	>82
Water	29,881[a]	-	31,488[a]	-	31,490[a]	-	426.7[a]	-	-	-	-	
Total	29,900[b]	634	31,490[b]	47.8	31,490[b]	<2.4	427.6[b]	2,032	15.4[b]	1,070,000[c]	92.1	>95

[a]Balance of total flow after accounting for organics
[b]Calculated from material balances using measured flows and concentrations
[c]Due to accuracy of concentration measurements
[d]Sample calculation: 505 ppmw × 29,900 kg/hr × $(10^6$ ppmw$)^{-1}$ = 15.1 kg/hr
[e]Values represent minimum removal efficiencies due to component concentrations below analytical detection limits
[f]Not used for calculation of removal efficiencies because of the need to calculate actual amount of organic removed from the waste stream and to minimize background interference effects.
[g]Calculated as: (1- (Effluent from stripper)/(Influent to stripper)) × 100%.

REFERENCES

Site A:

"Field Test and Evaluation of the Steam Stripping Process at B.F. Goodrich, La Porte, TX". Research Triangle Institute. Prepared for the U.S. EPA, Cincinnati, Ohio. EPA Contract No. 68-03-3253, December 5, 1986.

Site B:

"Post Sampling Report for Olin Chemical, Rochester, New York." Metcalf and Eddy. Prepared for the U.S. EPA, Cincinnati, Ohio. September, 1986.

Plant G:

"Hazardous Waste Pretreatment for Emission Control: Field Tests of Steam Stripping/Carbon Adsorption at Plant G," Research Triangle Institute, EPA Contract No. 68-02-3253, Work Assignment No. 5, August 1986.

Appendix E—Draft EPA Reference Methods 25D and 25E

METHOD 25D--DETERMINATION OF THE VOLATILE ORGANIC CONCENTRATION OF WASTE SAMPLES

Introduction

Performance of this method should not be attempted by persons unfamiliar with the operation of a flame ionization detector (FID) or a electrolytic conductivity detector (ELCD) because knowledge beyond the scope of this presentation is required.

1. Applicability and Principle

1.1 Applicability. This method is applicable to the determination of the volatile organic concentration of hazardous waste.

1.2 Principle. A sample of waste is collected at a point before the waste is exposed to the atmosphere such as in an enclosed pipe or other closed system that is used to transfer the waste after generation to the first hazardous waste management unit. The sample is then heated and purged with a stream of nitrogen to separate the organic compounds. Part of the sample is analyzed for carbon content, as methane, with an FID, and part of the sample is analyzed for chlorine content, as chloride, with an ELCD. The volatile organic concentration is the sum of the carbon and chlorine content of the sample.

2. Apparatus

2.1 Sampling. The following equipment is required:

2.1.1 Static Mixer. Installed in-line or as a by-pass loop, sized so that the drop size of the dispersed phase is no greater than 1000 microns (μ). If the installation of the mixer is in a by-pass loop, then the entire waste stream shall be diverted through the mixer.

2.1.2 Tap. Installed no further than two pipe diameters downstream of the static mixer outlet.

2.1.3 Sampling Tube. Flexible Teflon, 0.25 in. ID. NOTE: Mention of names or specific products does not constitute endorsement by the Environmental Protection Agency.

2.1.4 Sample Container. Borosilicate glass or polyterafluoroethylene (PTFE), 15 to 50 ml, and a Teflon lined screw cap capable of forming an air tight seal.

2.1.5 Cooling Coil. Fabricated from 0.25 in. ID 304 stainless steel tubing with a thermocouple at the coil outlet.

2.2 Analysis. The following equipment is required:

2.2.1 Purging Apparatus. For separating the organic compounds from the waste sample. A schematic of the system is shown in Figure 1. The purging apparatus consists of the following major components.

2.2.1.1 Purging Chamber. A glass container to hold the sample while it is heated and purged with dry nitrogen. Exact dimensions are shown in Figure 2.

The cap of the purging chamber is equipped with three fittings: one for a mechanical stirrer (fitting with the #11 Ace thread), one for a thermometer (top fitting), and one for the Teflon exit tubing (side fitting) as shown in Figure 2.

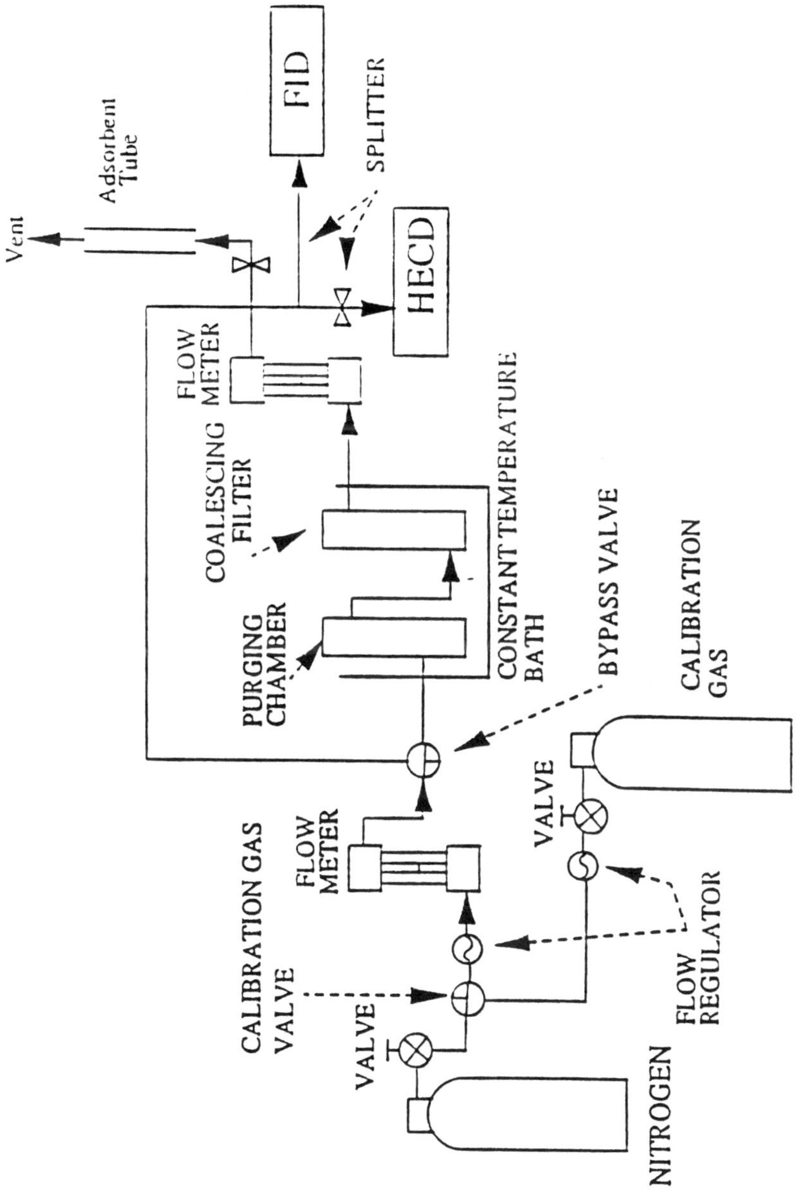
Vent
Adsorbent Tube
FID
SPLITTER
HECD
FLOW METER
COALESCING FILTER
CONSTANT TEMPERATURE BATH
PURGING CHAMBER
BYPASS VALVE
CALIBRATION GAS
VALVE
FLOW METER
CALIBRATION GAS VALVE
VALVE
FLOW REGULATOR
NITROGEN

FIGURE 1

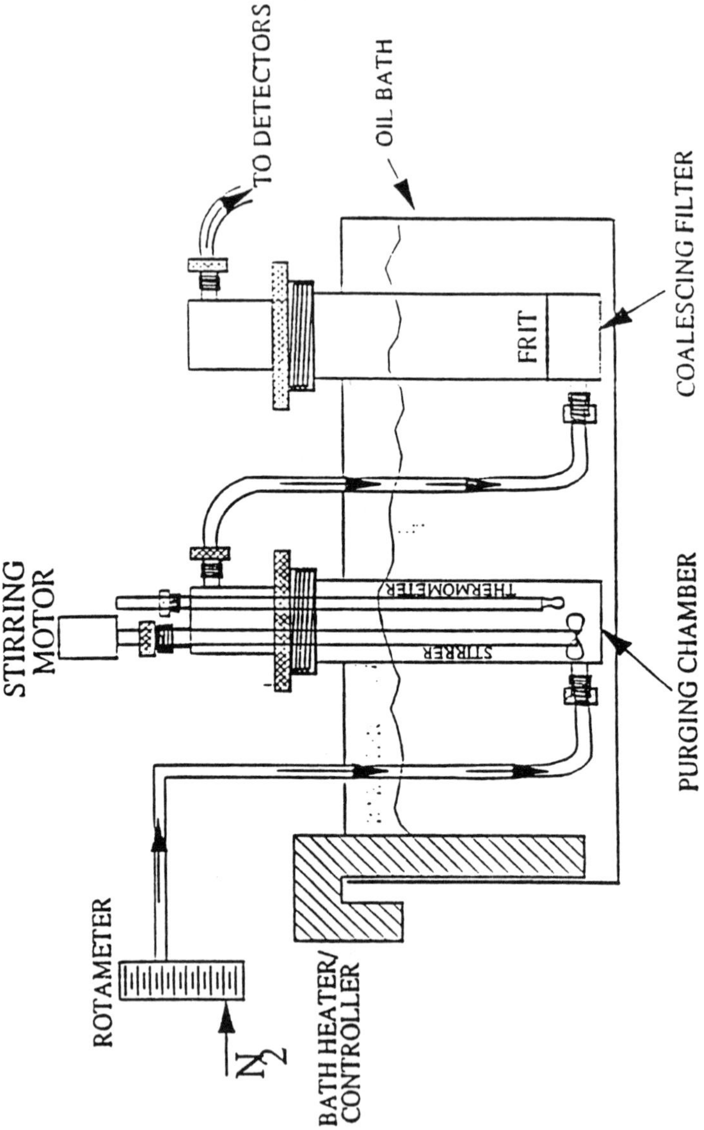
TO DETECTORS
OIL BATH
STIRRING MOTOR
FRIT
THERMOMETER
STIRRER
COALESCING FILTER
PURGING CHAMBER
ROTAMETER
N2
BATH HEATER/ CONTROLLER

FIGURE 2

The base of the purging chamber is a 50-mm inside diameter (ID) cylindrical glass tube. One end of the tube is fitted with a 50-mm Ace-thread fitting while the other end is sealed. Near the sealed end in the side wall is a fitting for a glass purging lance.

2.2.1.2 Purging Lance. Glass tube, 6-mm ID by 15.25 cm long, bent into an el shape. The el end of the tube is sealed and then pierced with fifteen holes each 1 mm in diameter.

2.2.1.3 Mechanical Stirrer. Stainless steel or PTFE stirring rod driven by an electric motor.

2.2.1.4 Coalescing Filter. Porous fritted disc incorporated into a container with the same dimensions as the purging chamber. The details of the design are shown in Figure 3.

2.2.1.5 Constant Temperature Bath. Capable of maintaining a temperature around the purging chamber and coalescing filter of 75 + 5°C.

2.2.1.6 Three-way valves. Two, manually operated, stainless steel.

2.2.1.7 Flow Controller. Capable of maintaining a purge gas flow rate of 6 + .006 l/min.

2.2.1.8 Rotameters. Two for monitoring the air flow through the purging system (0-20 l/min).

2.2.1.9 Sample Splitters. Two heated flow restrictors. At a purge rate of up to 6 l/min, one will supply a constant flow of 70 to 100 ml/min to the analyzers. The second will split the analytical flow between the FID and the ELCD. The approximate flow to the FID will be 40 ml/min and to the ELCD will be 15 ml/min, but the exact flow shall

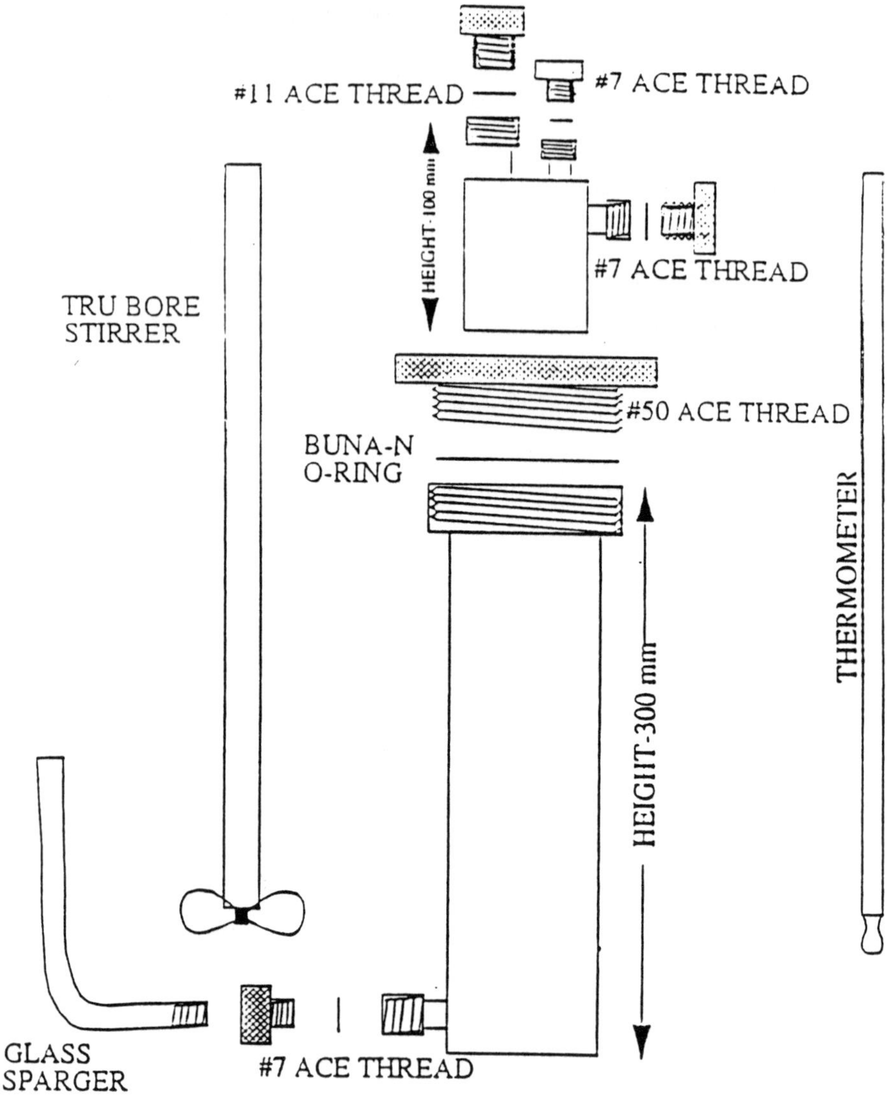
#11 ACE THREAD
#7 ACE THREAD
HEIGHT-100 mm
#7 ACE THREAD
TRU BORE
STIRRER
#50 ACE THREAD
BUNA-N
O-RING
THERMOMETER
HEIGHT-300 mm
GLASS
SPARGER
#7 ACE THREAD

FIGURE 3

be adjusted to be compatible with the individual detector and to meet its linearity requirement.

2.2.1.10 Adsorbent Tube. To hold 10 g of activated charcoal. Excess purge gas is vented through the adsorbent tube to prevent any potentially hazardous materials from entering the laboratory.

2.2.2 Volatile Organic Measurement System. Consisting of an FID to measure the carbon content, as methane, of the sample and an ELCD to measure the chlorine content.

2.2.2.1 FID. An FID meeting the following specifications is required:

2.2.2.1.1 Linearity. A linear response (+ 5 percent) over the operating range as demonstrated by the procedures established in Section 5.1.1.

2.2.2.1.2 Range. A full scale range of 50 pg carbon/sec to 50 μg carbon/sec. Signal attenuators shall be available to produce a minimum signal response of 10 percent of full scale.

2.2.2.1.3 Data Recording System. Analog strip chart recorder or digital integration system compatible with the FID for permanently recording the output of the detector.

2.2.2.2 ELCD. An ELCD meeting the following specifications is required:

2.2.2.2.1 Linearity. A linear response (+ 10 percent) over the response range as demonstrated by the procedures in Section 5.1.2

2.2.2.2.2 Range. A full scale range of 5.0 pg/sec to 500 ng/sec chloride. Signal attenuators shall be available to produce a minimum signal response of 10 percent of full scale.

2.2.2.2.3 Data Recording System. Analog strip chart recorder or digital integration system compatible with the output voltage range of the ELCD.

3. Reagents

3.1 Sampling.

3.1.1 Polyethylene Glycol (PEG). 98 percent pure with an average molecular weight of 400. Remove any organic compounds already present in the PEG before it is used by heating it to 250°C and purging it with nitrogen at a flow rate of 1 to 2 l/min for 2 hours.

3.2 Analysis.

3.2.1 Sample Separation. The following are required for the sample purging step:

3.2.1.1 PEG. Same as Section 3.1.1.

3.2.1.2 Silicon, Mineral, or Peanut Oil. For use as the heat dispersing medium in the constant temperature bath.

3.2.1.3 Purging Gas. Zero grade nitrogen (N_2), containing less than 1 ppm carbon.

3.2.2 Volatile Organic Measurement. The following are required for measuring the volatile organic concentration:

3.2.2.1 Hydrogen (H_2). Zero grade H_2, 99.999 percent pure.

3.2.2.2 Combustion Gas. Zero grade air or oxygen as required by the FID.

3.2.2.3 FID Calibration Gases.

3.2.2.3.1 Low-level Calibration Gas. Gas mixture standard with a nominal concentration of 35 ppmv propane in N_2.

3.2.2.3.2 Mid-level Calibration Gas. Gas mixture standard with a nominal concentration of 175 ppmv propane in N_2.

3.2.2.3.3 High-level Calibration Gas. Gas mixture standard with a nominal concentration of 350 ppmv propane in N_2.

3.2.2.4 ELCD Calibration Gases.

3.2.2.4.1 Low-level Calibration Gas. Gas mixture standard with a nominal concentration of 20 ppmv 1,1-dichloroethene in N_2.

3.2.2.4.2 Mid-level Calibration Gas. Gas mixture standard with a nominal concentration of 100 ppmv 1,1-dichloroethene in N_2.

3.2.2.4.3 High-level Calibration Gas. Gas mixture standard with a nominal concentration of 200 ppmv 1,1-dichloroethene in N_2.

3.2.2.5 Water. Deionized distilled water that conforms to American Society for Testing and Materials Specification D 1193-74, Type 3, is required for analysis. At the option of the analyst the $KMnO_4$, test for oxidizable organic matter may be omitted when high concentrations are not expected to be present.

3.2.2.6 N-Propanol. ACS grade or better.

3.2.2.7 Electrolyte Solution. For use in the conductivity detector. Mix together 500 ml of water and 500 ml of n-propanol and store in a glass container.

3.2.2.8 Charcoal. Activated coconut, 12 to 30 mesh.

4.0 Procedure

4.1 Sampling.

4.1.1 Sampling Plan Design and Development. Use the procedures in chapter nine of the Office of Solid Waste's publication, Test Methods for Evaluating Solid Waste, third edition (SW-846), as guidance in developing a sampling plan.

4.1.2 Waste in Enclosed Pipes. Sample at one of the locations described in Section 1.2 in order to minimize the loss of organic compounds. Assemble the sampling apparatus as shown in Figure 4. Install the static mixer in the process line or in a by-pass line. Locate the tap within two pipe diameters of the static mixer outlet.

Prepare the sampling containers as follows: Pour into the container an amount of PEG equal to the total volume of the sample container minus 10 ml. PEG will reduce but not eliminate the loss of organic compounds during sample collection. Weigh the sample container with the screw cap, the PEG, and any labels to the nearest 0.01 g and record the weight (m_{st}). Before sampling, store the containers in an ice bath until the temperature of the PEG is less than 4°C.

Begin sampling by purging the sample lines and cooling coil with at least four volumes of waste. Collect the purged material in a separate container and dispose of it properly.

After purging, stop the sample flow and direct the sampling tube to a preweighed sample container, prepared as previously described of this section. Keep the tip of the tube below the surface of the PEG during sampling to minimize contact with the atmosphere. Sample at a flow rate such that the temperature of the waste is less than 10°C. Fill the sample container and immediately cap it (within 5 seconds) so that a minimum headspace exists in the container. Store immediately in a cooler and cover with ice.

Alternative sampling techniques may be used upon the approval of the Administrator.

4.2 Sample Recovery. Remove the sample container from the cooler, and wipe the exterior of the container to remove any extraneous ice, water, or other debris. Reweigh the sample container and sample to the nearest 0.01 g, and record the weight (m_{sf}). Pour the contents of the

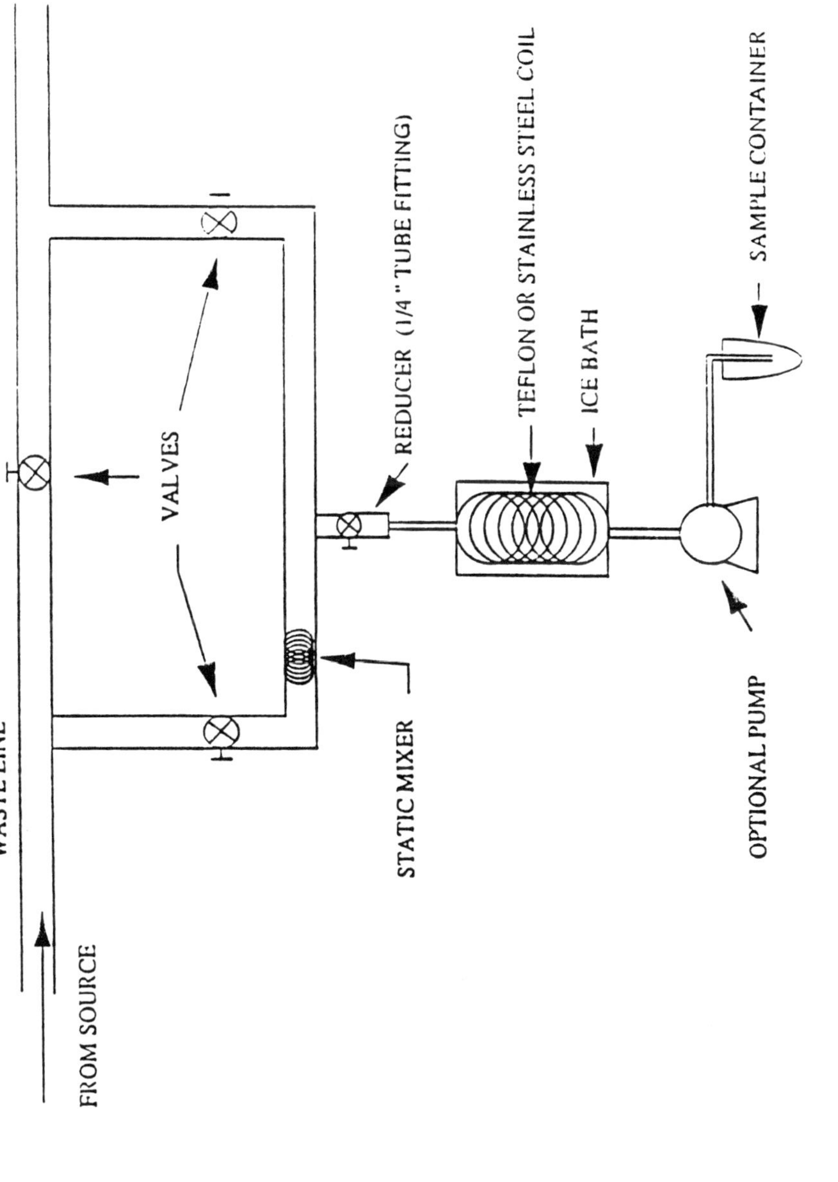
WASTE LINE
FROM SOURCE
VALVES
STATIC MIXER
REDUCER (1/4" TUBE FITTING)
TEFLON OR STAINLESS STEEL COIL
ICE BATH
OPTIONAL PUMP
SAMPLE CONTAINER

FIGURE 4

sample container into the purging flask, rinse the sample container three times with PEG, transferring the rinsings to the purging flask after each rinse. The total volume of PEG in the purging flask shall be approximately 50 ml. Add approximately 50 ml of water. Assemble the purging apparatus as shown in Figure 1, leaving the purging chamber out of the constant temperature bath. Adjust the stirring rod so that it nearly reaches the bottom of the chamber. Position the sparger so that it is within 1 cm of the bottom but does not interfere with the stirring rod. Lower the thermometer so that it extends into the liquid.

4.3 Sample Analysis. Turn on the constant temperature bath and allow the temperature to equilibrate at 75 + 5°C. Turn the bypass valve so that the purge gas bypasses the purging chamber. Turn on the purge gas. Allow both the FID and the ELCD to warm up until a stable baseline is achieved on each detector. Pack the adsorbent tube with 10 g of charcoal. Change this after each run and dispose of the spent charcoal properly. Place the assembled chamber in the constant temperature bath. When the temperature of the liquid reaches 75 + 5°C, turn the bypass valve so that the purge gas flows through the purging chamber. Begin recording the response of the FID and the ELCD. Compare the readings between the two rotameters in the system. If the readings differ by more than five percent, stop the purging and determine the source of the discrepancy before resuming.

As the purging continues, monitor the output of the FID to make certain that the separation is proceeding correctly and that the results are being properly recorded. Every 10 minutes read and record the purge flow rate and the liquid temperature. Continue the purging for 30 minutes.

4.4 Water Blank. Transfer about 60 ml of water into the purging chamber. Add 50 ml of PEG to the purging chamber. Treat the blank as described in Sections 4.2 and 4.3 beginning with the fifth sentence of Section 4.2.

5. Operational Checks and Calibration

Maintain a record of performance of each item.

5.1 Initial Performance Check of Purging System. Before placing the system in operation, after a shutdown of greater than six months, and after any major modifications, conduct the linearity checks described in Sections 5.1.1 and 5.1.2. Install all calibration gases at the three-way calibration gas valve. See Figure 1.

5.1.1 FID Linearity Check and Calibration. With the purging system operating as in Section 4.3, allow the FID to establish a stable baseline. Set the secondary pressure regulator of the calibration gas cylinder to the same pressure as the purge gas cylinder and inject the calibration gas by turning the calibration gas valve, to switch flow from the purge gas to the calibration gas. Continue the calibration gas flow for approximately two minutes before switching to the purge gas. Make triplicate injections of each calibration gas (Section 3.2.2.3), and then calculate the average response factor for each concentration (R_t), as well as the overall mean of the response factor values, R_o. The instrument linearity is acceptable if each R_t is within 5 percent of R_o and if the relative standard deviation (Section 6.9) for each set of triplicate injections is less than 5 percent.

Record the overall mean value of the propane response factor values as the FID calibration response factor, R_o.

5.1.2 ELCD Linearity Check and Calibration. With the purging system operating as in Section 4.3, allow the ELCD to establish a stable baseline. Set the secondary pressure regulator of the calibration gas cylinder to the same pressure as the purge gas cylinder and inject the calibration gas by turning the calibration gas valve to switch flow from the purge gas to the calibration gas. Continue the calibration gas flow about two minutes before switching to the purge gas. Make triplicate injections of each calibration gas (Section 3.2.2.4), and then calculate the average response factor for each concentration, R_{th}, as well as the overall mean of the response factors, R_{oh}. The instrument linearity is acceptable if each Rth is within 10 percent of R_{oh} and if the relative standard deviation (Section 6.9) for each set of triplicate injections is less than 10 percent. Record the overall mean value of the chlorine response factors as the ELCD response factor, R_{oh}.

5.2 Daily Calibrations.

5.2.1 FID Daily Calibration. Inject duplicate samples from the mid- level FID calibration gas (Section 3.2.2.3.2) as described in Section 5.1.1, and calculate the average daily response factor (DR_t). System operation is adequate if the DR_t is within 5 percent of the R_o calculated during the initial performance test (Section 5.1.1). Use the DR_t for calculation of carbon content in the samples.

5.2.2 ELCD Daily Calibration. Inject duplicate samples from the mid-level ELCD calibration gas (Section 3.2.2.4.2) as described in Section 5.1.2 and calculate the average daily response factor DR_{th}.

The system operation is adequate if the DR_{th} is within 10 percent of the R_{oh} calculated during the initial performance test (Section 5.1.2). Use the DR_{th} for calculation of chlorine in the samples.

5.3 Analytical Balance. Calibrate against standard weights.

6.0 Calculations

6.1 Nomenclature.

A_b = Area under the water blank response curve, counts.

A_s = Area under the sample response curve, counts.

C = Concentration of volatile organics in the sample, ppmw.

C_c = Concentration of FID calibration gas, ppmv.

C_h = Concentration of ELCD calibration gas, ppmv.

DR_t = Average daily response factor of the FID, μg C/counts.

DR_{th} = Average daily response factor of the ELCD, μg Cl-/counts.

m_{co} = Mass of carbon, as methane, in the FID calibration standard, μg.

m_{ch} = Mass of chloride in the ELCD calibration standard, μg.

m_s = Mass of the waste sample, g.

m_{sc} = Mass of carbon, as methane, in the sample, μg.

m_{sf} = Mass of sample container and waste sample, g.

m_{sh} = Mass of chloride in the sample, μg.

m_{st} = Mass of sample container prior to sampling, g.

m_{vo} = Mass of volatile organics in the sample, μg.

P_a = Ambient barometric pressure in the laboratory, Torr.

Q_c = Flowrate of calibration gas, l/min.

t_c = Length of time standard gas is delivered to the analyzer, min.

T_a = Ambient temperature in the laboratory, °K.

6.2 Mass of Carbon, as Methane, in the FID Calibration Gas.

$$m_{co} = k_2 C_c t_c Q_c (P_a/T_a) \quad \text{Eq. 1}$$

where $k_2 = 0.7694 \times 10\text{-}3$ µg C-°K/ PPMV-Torr

6.3 Mass of Chloride in the ELCD Calibration Gas.

$$m_{ch} = k_3 C_h t_c Q_c (P_a/T_a) \quad \text{Eq. 2}$$

where $k_3 = 1.1371 \times 10\text{-}3$ µg Cl-°K/ PPMV-Torr

6.4 FID Response Factor.

$$R_t = m_{co}/A \quad \text{Eq.3}$$

6.5 ELCD Response Factor.

$$R_{th} = m_{ch}/A \quad \text{Eq. 4}$$

6.6 Mass of Carbon, as Methane, in the Sample.

$$m_{sc} = DR_t (A_s - A_b) \quad \text{Eq. 5}$$

6.7 Mass of Chloride in the Sample.

$$m_{sh} = DR_{th} (A_s - A_b) \quad \text{Eq. 6}$$

6.8 Mass of Volatile Organics in the Sample.

$$m = m_{sc} + m_{sh} \quad \text{Eq. 7}$$

6.9 Relative Standard Deviation.

$$RSD = \text{-}100/x \left[\sum_{i=1}^{n}\text{-}(xi\text{-}x)2/(n\text{-}1)\right]1/2 \quad \text{Eq. 8}$$

6.10 Mass of Sample.

$$m_s = m_{sf} - m_{st} \qquad \text{Eq. 9}$$

6.11 Concentration of Volatile Organics in Waste.

$$C = m_{vo}/m_s \qquad \text{Eq.10}$$

METHOD 25E--DETERMINATION OF VAPOR PHASE ORGANIC CONCENTRATION IN WASTE SAMPLES

Introduction

Performance of this method should not be attempted by persons unfamiliar with the operation of a flame ionization detector (FID) nor by those who are unfamiliar with source sampling because knowledge beyond the scope of this presentation is required.

1. Applicability and Principle

1.1 Applicability. This method is applicable for determining the vapor pressure of waste samples from treatment, storage, and disposal facilities (TSDF).

1.2 Principle. A waste sample is collected from a source just prior to entering a tank. The headspace vapor of the sample is analyzed for carbon content by a headspace analyzer, which uses an FID.

2. Interferences

2.1 The analyst shall select the operating parameters best suited to his requirements for a particular analysis. The analyst shall produce confirming data through an adequate supplemental analytical technique and have the data available for review by the Administrator.

3. Apparatus

3.1 Sampling. The following equipment is required:

3.1.1 Sample Containers. Vials, glass, with butyl rubber septa, Perkin-Elmer Corporation Numbers 0105-0129 (glass vials), B001-0728 (gray butyl rubber septum, plug style), 0105-0131 (butyl rubber septa), or equivalent. The seal shall be made from butyl rubber. Silicone rubber seals are not acceptable.

3.1.2 Vial Sealer. Perkin-Elmer Number 105-0106, or equivalent.

3.1.3 Gas-Tight Syringe. Perkin-Elmer Number 00230117, or equivalent. pipe:

3.1.4.1 Static mixer. In-line or by-pass loop, sized so that the drop size of the dispersed phase is no greater than 1000 μm. If the mixer is installed as a by-pass loop, the entire waste stream shall be diverted through the mixer.

3.1.4.2 Tap.

3.1.4.3 Tubing. Teflon, 0.25-in. ID. Note: Mention of trade names or specific products does not constitute endorsement by the Environmental Protection Agency.

3.1.4.4 Cooling Coil. Stainless steel (304), 0.25 in.-ID, equipped with a thermocouple at the coil outlet.

3.2 Analysis. The following equipment is required:

3.2.1 Balanced Pressure Headspace Sampler. Perkin-Elmer HS-6, HS-100, or equivalent, equipped with a glass bead column instead of a chromatographic column.

3.2.2 FID. An FID meeting the following specifications is required:

3.2.2.1 Linearity. A linear response (±5 percent) over the operating range as demonstrated by the procedures established in Section 6.1.2.

3.2.2.2 Range. A full scale range of 1 to 10,000 ppm CH_4. Signal attenuators shall be available to produce a minimum signal response of 10 percent of full scale.

3.2.3 Data Recording System. Analog strip chart recorder or digital integration system compatible with the FID for permanently recording the output of the detector.

3.2.4 Thermometer. Capable of reading temperatures in the range of 30° to 60°C with an accuracy of ±0.1°C.

4. Reagents

4.1 Analysis. The following items are required for analysis:

4.1.1 Hydrogen (H_2). Zero grade.

4.1.2 Carrier Gas. Zero grade nitrogen, containing less than 1 ppm carbon (C) and less than 1 ppm carbon dioxide.

4.1.3 Combustion Gas. Zero grade air or oxygen as required by the FID.

4.2 Calibration and Linearity Check.

4.2.1 Stock Cylinder Gas Standard. 100 percent propane. The manufacturer shall: (a) certify the gas composition to be accurate to ±3 percent or better (see Section 4.2.1.1); (b) recommend a maximum shelf life over which the gas concentration does not change by greater than ±5 percent from the certified value; and (c) affix the date of gas cylinder preparation, certified propane concentration, and recommended maximum shelf life to the cylinder before shipment to the buyer.

4.2.1.1 Cylinder Standards Certification. The manufacturer shall certify the concentration of the calibration gas in the cylinder by (a) directly analyzing the cylinder and (b) calibrating his analytical procedure on the day of cylinder analysis. To calibrate his analytical procedure, the manufacturer shall use, as a minimum, a three-point calibration curve.

4.2.1.2 Verification of Manufacturer's Calibration Standards. Before using, the manufacturer shall verify each calibration standard by (a) comparing it to gas mixtures prepared in accordance with the procedure described in Section 7.1 of Method 106 of Part 61, Appendix B, or by (b) calibrating it against Standard Reference Materials (SRM's) prepared by the National Bureau of Standards, if such SRM's are available. The agreement between the initially determined concentration value and the verification concentration value shall be within ±5 percent. The manufacturer shall reverify all calibration standards on a time interval consistent with the shelf life of the cylinder standards sold.

5. Procedure

5.1 Sampling.

5.1.1 Sampling Plan Design and Development. Use the procedures in chapter nine of the Office of Solid Waste's publication, Test Methods for Evaluating Solid Waste, third edition (SW-846), as guidance in developing a sampling plan.

5.1.2 Sample according to the procedures in chapter nine of SW-846, or, if sampling from an enclosed pipe, sample according to the procedures described below.

5.1.2.1 The sampling apparatus designed to sample from an enclosed pipe is shown in Figure 1, and consists of an in-line static mixer, a tap, a cooling coil immersed in an ice bath, a flexible Teflon tube connected to the outlet of the cooling coil, and a sample container.

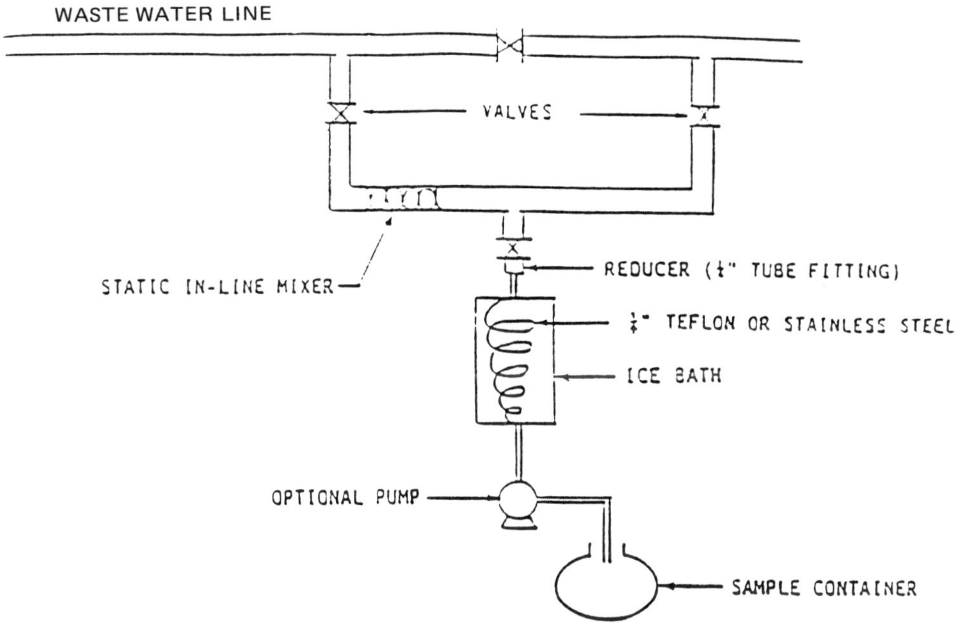

Figure 1. Schematic of sampler equipped with static mixer and cooling coil.

Locate the tap within two pipe diameters of the static mixer outlet. Install the static mixer in the process line or in a by-pass line.

5.1.2.2 Begin sampling by purging the sample lines and cooling coil with at least four volumes of waste. Collect the purged material in a separate container. Consider the purged material hazardous waste and dispose of it properly.

5.1.2.3 After purging, stop the sample flow and transfer the Teflon sampling tube to a sample container. Sample at a flow rate such that the temperature of the waste is <10°C (<50°F). Fill the sample container halfway (±5 percent) and cap it within 5 seconds.

5.1.2.4 Store the collected samples in ice or a refrigerator until analysis.

5.1.2.5 Alternative sampling techniques may be used upon the approval of the Administrator.

5.2 Analysis.

5.2.1 Allow one hour for the headspace vials to equilibrate at the temperature specified in the regulation. Allow the FID to warm up.

5.2.2 Check the calibration of the FID daily using the procedures in Section 6.1.2.

5.2.3 Follow the manufacturer's recommended procedures for the normal operation of the headspace sampler and FID.

5.2.4 Use the procedures in Sections 7.4 and 7.5 to calculate the vapor phase organic vapor pressure in the samples.

5.2.5 Monitor the output of the detector to make certain that the results are being properly recorded.

6. Operational Checks and Calibration.

Maintain a record of performance of each item.

6.1 Use the procedures in Sections 6.1.1 to calibrate the headspace analyzer and FID and check for linearity before the system is first placed in operation, after any shutdown longer than 6 months, and after any modification of the system.

6.1.1 Calibration and Linearity. Use the procedures in Section 6.2.1 of Method 18 of Part 60, Appendix A, to prepare the standards and calibrate the flowmeters, using propane as the standard gas. Fill the calibration standard vials halfway (±5 percent) with deionized water. Purge and fill the airspace calibration standards in triplicate at concentrations that will bracket the applicable cutoff. For a cutoff of 5.2 kPa (0.75 psi), prepare nominal concentrations of 30,000, 50,000, and 70,000 ppm as propane. For a cutoff of 27.6 kPa (4.0 psi), prepare nominal concentrations of 200,000, 300,000, and 400,000 ppm as propane.

6.1.1.1 Use the procedures in Section 5.2.3 to measure the FID response of each standard. Use a linear regression analysis to calculate the values for the slope (k) and the y-intercept (b). Use the procedures in Sections 7.2 and 7.3 to test the calibration and the linearity.

6.1.2 Daily FID Calibration Check. Check the calibration at the beginning and at the end of the daily runs by using the following procedures. Prepare two calibration standards at the nominal cutoff concentration using the procedures in Section 6.1.1. Place one at the beginning and one at the end of the daily run. Measure the FID response of the daily calibration standard and use the values for k and b from the most recent calibration to calculate the concentration of the daily standard. Use an equation similar to 25E-2 to calculate the percent difference between the daily standard and C_s. If the difference is within 5 percent, then the previous values for k and b may be used. Otherwise, use the procedures in Section 6.1.1 to recalibrate the FID.

7. Calculations

7.1 Nomenclature.

A = Measurement of the area under the response curve, counts.

b = y-intercept of the linear regression line.

C_a = Measured vapor phase organic concentration of sample, ppm as propane.

C_{ma} = Average measured vapor phase organic concentration of standard, ppm as propane.

C_m = Measured vapor phase organic concentration of standard, ppm as propane.

C_s = Calculated standard concentration, ppm as propane.

k = Slope of the linear regression line.

P_{bar} = Atmospheric pressure at analysis conditions, mm Hg (in. Hg).

P^* = Organic vapor pressure in the sample, kPa (psi).

β = 1.333 X 10-7 kPa/[(mm Hg)(ppm)], (4.91 X 10-7 psi/[(in. Hg)(ppm)])

7.2 Linearity. Use the following equation to calculate the measured standard concentration for each standard vial.

$$C_m = k A + b \qquad \text{Eq. 25E-1}$$

7.2.1 Calculate the average measured standard concentration (C_{ma}) for each set of triplicate standards and use the following equation to calculate the percent difference between C_{ma} and C_s.

$$\text{Percent Difference} = \frac{C_s - C_{ma}}{C_s} \times 100 \qquad \text{Eq. 25E-2}$$

The instrument linearity is acceptable if the percent difference is within five for each standard.

7.3. Relative Standard Deviation (RSD). Use the following equation to calculate the RSD for each triplicate set of standards.

$$RSD = \frac{100}{C_{ma}} \quad \frac{\Sigma\,(C_m - C_{ma})2}{n-1} \qquad \text{Eq. 25E-3}$$

The calibration is acceptable if the RSD is within five for each standard concentration.

7.4 Concentration of organics in the headspace. Use the following equation to calculate the concentration of vapor phase organics in each sample.

$$C_a = k A + b \qquad \text{Eq. 25E-4}$$

7.5 Vapor Pressure of Organics in the Headspace Sample. Use the following equation to calculate the vapor pressure of organics in the sample.

$$P^* = \beta \; Pbar \; C_a \qquad \text{Eq. 25E-5}$$

Appendix F—Example Facility Analysis

F.1 INTRODUCTION

The purpose of this document is to provide technical information to States on best available control technology (BACT) and lowest achievable emission rate (LAER) determinations for controlling emissions of VOC from industrial wastewaters. Steam stripping is an effective control technology, however, other controls which achieve the same performance can be applied. To determine which wastewater streams should be controlled, the impacts of applying various control options to a set of example facilities that collect and treat industrial wastewater were evaluated.

The purpose of this Appendix is to present the background information to support the development of a set of example facilities. Section F.2 presents a description of the example facilities, and Section F.3 presents the method used to predict the VOC emissions from these facilities. The methods used to predict steam stripper removal and test method concentration are presented in Sections F.4 and F.5, respectively. Section F.6 presents a brief discussion on hazardous air pollutants.

F.2 OVERVIEW OF EXAMPLE FACILITIES

The example facilities were developed from responses to questionnaires sent out by the Office of Air Quality Planning and Standards (OAQPS) in the Organic Chemicals, Plastics, and Synthetic Fibers (OCPSF) industry under the authority of Section 114 of the Clean Air Act. This information was supplemented with information gathered by the Office of Water Regulations and Standards (OWRS) under Section 308 of the Clean Water Act and with information collected by OAQPS during site visits.

Six example facilities were developed from this information. These facilities are examples of wastewater collection and treatment systems found in the targeted industries discussed in Chapter 2 of this document. Tables F-1 through F-6 present the characteristics of the individual wastewater streams for each of these example facilities. Figures F-1 through F-6 show the treatment and collection system configurations for each example facility.

TABLE F-1. EXAMPLE FACILITY 1 WASTEWATER STREAM CHARACTERISTICS

Stream ID	Flow (lpm)	Organic Concentration (mg/l)	VOC Test Method Concentration (mg/l)	Fraction Emitted (wt%)	Uncontrolled VOC Emissions (Mg/yr)	Controlled VOC Emissions (Mg/yr)	VOC Emission Reduction (Mg/yr)	VOC Emission Reduction (wt%)
1	28	11,995.0	1,740.1	18.26	32.69	27.90	4.7876	14.64
2	568	1,250.0	1,289.0	75.10	280.22	0.04	280.1816	99.99
3	568	9,935.0	10,325.8	39.61	1174.73	0.17	1174.5520	99.99
4	118	6,520.0	6,727.9	63.59	257.81	0.04	257.7653	99.98
5	101	12,000.0	5,280.0	11.52	73.26	46.09	27.1710	37.09
6	909	364.0	443.3	45.53	79.16	0.01	79.1460	99.99
7	663	414.0	526.1	45.11	68.7956	0.00	68.7956	100.00
8	568	59.0	71.1	61.18	10.77	0.04	10.7364	99.65
9	1,278	191.0	87.5	16.35	20.98	13.62	7.3538	35.06
	------				------	------	------	------
	4,800				1,998	88	1,910	76

TABLE F-2. EXAMPLE FACILITY 2 WASTEWATER STREAM CHARACTERISTICS

Stream ID	Flow (lpm)	Organic Concentration (mg/l)	VOC Test Method Concentration (mg/l)	Fraction Emitted (wt%)	Uncontrolled VOC Emissions (Mg/yr)	Controlled VOC Emissions (Mg/yr)	VOC Emission Reduction (Mg/yr)	VOC Emission Reduction (wt%)
1	0.08	5.0	6.1	92.20	0.00	0.00	0.0002	106.91
2	0.003	8.4	0.6	98.84	0.00	0.00	0.0000	1.22
3	0.002	0.4	0.5	92.50	0.00	0.00	0.0000	106.60
4	0.003	25.7	31.3	91.46	0.00	0.00	0.0000	107.25
5	0.10	0.5	0.6	92.50	0.00	0.00	0.0000	106.48
6	0.004	0.9	1.1	100.00	0.00	0.00	0.0000	92.02
	------				------	------	------	------
	0				0	0	0	87

TABLE F-3. EXAMPLE FACILITY 3 WASTEWATER STREAM CHARACTERISTICS

Stream ID	Flow (lpm)	Organic Concentration (mg/l)	VOC Test Method Concentration (mg/l)	Fraction Emitted (wt%)	Uncontrolled VOC Emissions (Mg/yr)	Controlled VOC Emissions (Mg/yr)	VOC Emission Reduction (Mg/yr)	VOC Emission Reduction (wt%)
1	47	700.0	854.0	56.93	9.91	0.00	9.9111	99.99
2	45	1.0	1.2	96.67	0.02	0.00	0.0228	99.99
3	6	460.0	563.8	45.41	0.62	0.00	0.6234	99.99
4	31	1,460.0	1,655.8	63.55	14.98	0.00	14.9826	99.99
5	710	101.0	123.2	41.78	15.75	0.00	15.7430	99.99
6	454	80.0	97.4	41.85	7.99	0.00	7.9935	99.99
7	38	300.0	315.0	51.46	3.07	0.00	3.0714	99.99
8	950	1,500.4	1,067.1	1.90	14.20	8.53	5.6713	39.93
9	685	49.0	51.0	65.26	11.52	0.00	11.5164	99.99
10	1,022	10,207.5	10,616.2	38.68	2121.17	0.31	2120.8610	99.99
11	449	1,005.0	1,046.5	41.63	98.65	0.01	98.6400	99.99
12	175	611.0	744.9	53.48	30.07	0.00	30.0672	99.99
13	568	222.7	269.0	54.77	36.40	0.01	36.3992	99.99
14	16	361.0	376.3	42.63	1.31	0.00	1.3114	99.99
	------				------	------	------	------
	5,196				2,366	9	2,357	96

TABLE F-4. EXAMPLE FACILITY 4 WASTEWATER STREAM CHARACTERISTICS

Stream ID	Flow (lpm)	Organic Concentration (mg/l)	VOC Test Method Concentration (mg/l)	Fraction Emitted (wt%)	Uncontrolled VOC Emissions (Mg/yr)	Controlled VOC Emissions (Mg/yr)	VOC Emission Reduction (Mg/yr)	VOC Emission Reduction (wt%)
1	76	800.0	976.0	61.82	19.68	0.00	19.6800	99.99
2	466	170.0	151.3	26.30	10.94	5.02	5.9227	54.13
3	5,300	60.0	50.4	31.14	52.05	15.08	36.9689	71.03
4	303	60.0	50.4	30.56	2.92	0.85	2.0737	71.03
5	719	60.0	50.4	32.40	7.35	2.13	5.2207	71.03
6	630	55.0	38.5	33.21	6.05	1.70	4.3467	71.90
7	0	250.0	192.5	35.29	0.01	0.00	0.0071	86.93
8	95	1,220.0	1,085.8	29.97	18.19	8.34	9.8446	54.13
9	176	840.0	722.4	36.07	28.03	4.05	23.9789	85.55
10	650	1,260.0	970.2	20.46	88.05	52.49	35.5582	40.38
11	60	110.0	77.7	32.73	1.13	0.30	0.8266	73.26
12	9	1,100.0	790.0	47.90	2.36	0.06	2.2985	97.43
13	10	200.0	158.0	34.36	0.36	0.04	0.3166	87.93
14	5	60.0	51.0	37.41	0.06	0.04	0.0125	21.78
15	6	800.0	648.0	32.16	0.77	0.16	0.6080	79.16
	------				------	------	------	------
	8,503				238	90	148	71

TABLE F-5. EXAMPLE FACILITY 5 WASTEWATER STREAM CHARACTERISTICS

Stream ID	Flow (lpm)	Organic Concentration (mg/l)	VOC Test Method Concentration (mg/l)	Fraction Emitted (wt%)	Uncontrolled VOC Emissions (Mg/yr)	Controlled VOC Emissions (Mg/yr)	VOC Emission Reduction (Mg/yr)	VOC Emission Reduction (wt%)
1	9	4,000.0	52.0	17.89	3.56	2.97	0.5941	16.69
2	9	1,100.0	14.3	17.89	0.89	0.74	0.1481	16.69
3	47	10,000.0	130.0	17.89	44.51	37.08	7.4267	16.69
4	47	20,000.0	18,800.0	75.22	374.19	0.06	374.1323	99.99
5	41	15,000.0	13,050.0	14.34	46.44	13.53	32.9178	70.88
6	11	234.0	203.6	14.34	0.20	0.06	0.1420	70.88
7	70	60,000.0	39,510.0	19.48	431.13	185.42	245.7092	56.99
8	11	242.0	159.2	19.39	0.28	0.12	0.1596	56.94
9	82	7,100.0	7,636.0	50.51	153.91	18.95	134.9550	87.69
10	607	58.0	47.7	35.63	6.59	5.17	1.4178	21.52
11	1	22,000.0	19,320.0	54.53	3.18	1.16	2.0254	63.63
12	1	10,000.0	7,700.0	2.80	0.14	0.08	0.0562	40.38
13	16	5,000.0	4,350.0	19.42	8.05	1.19	6.8571	85.19
14	6	5,000.0	3,950.0	21.54	3.61	0.44	3.1731	87.93
	------				------	------	------	------
	959				1,077	267	810	57

TABLE F-6. EXAMPLE FACILITY 6 WASTEWATER STREAM CHARACTERISTICS

Stream ID	Flow (lpm)	Organic Concentration (mg/l)	VOC Test Method Concentration (mg/l)	Fraction Emitted (wt%)	Uncontrolled VOC Emissions (Mg/yr)	Controlled VOC Emissions (Mg/yr)	VOC Emission Reduction (Mg/yr)	VOC Emission Reduction (wt%)
1	114	1,352.0	1,124.2	95.89	77.39	1.16	76.2347	98.51
2	189	3,258.0	2,574.0	75.34	244.22	21.38	222.8404	91.25
3	423	20.9	17.0	85.34	3.97	0.19	3.7741	95.09
4	151	1,268.0	1,065.1	100.00	100.93	0.01	100.9139	99.99
	------				------	------	------	------
	878				427	23	404	96

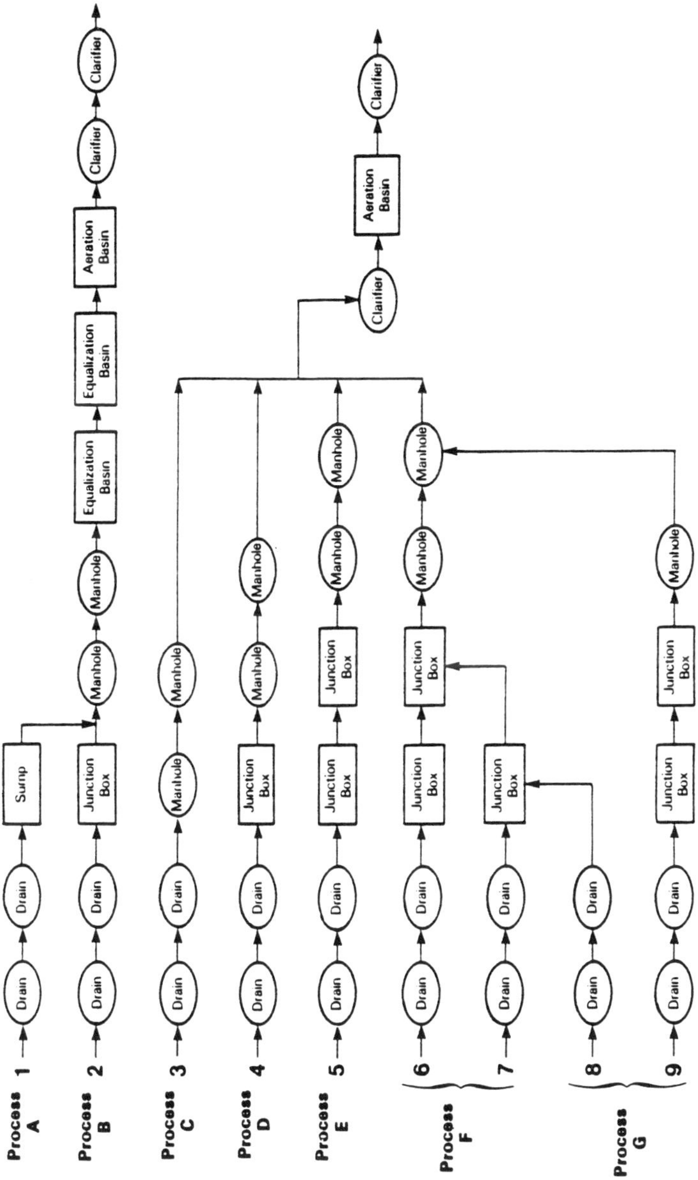

Figure F-1. Example Facility One.
Treatment Configuration and Collection System.

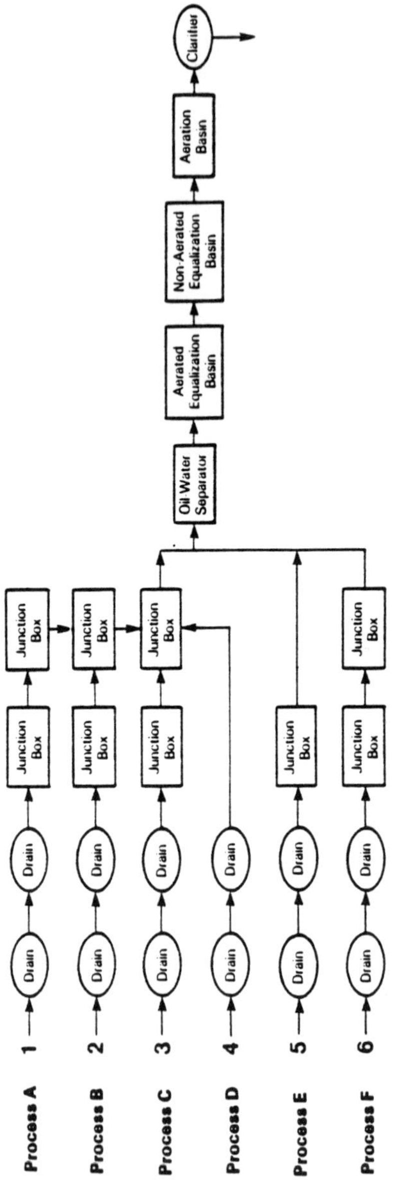

Figure F-2. Example Facility Two.
Treatment Configuration and Collection System.

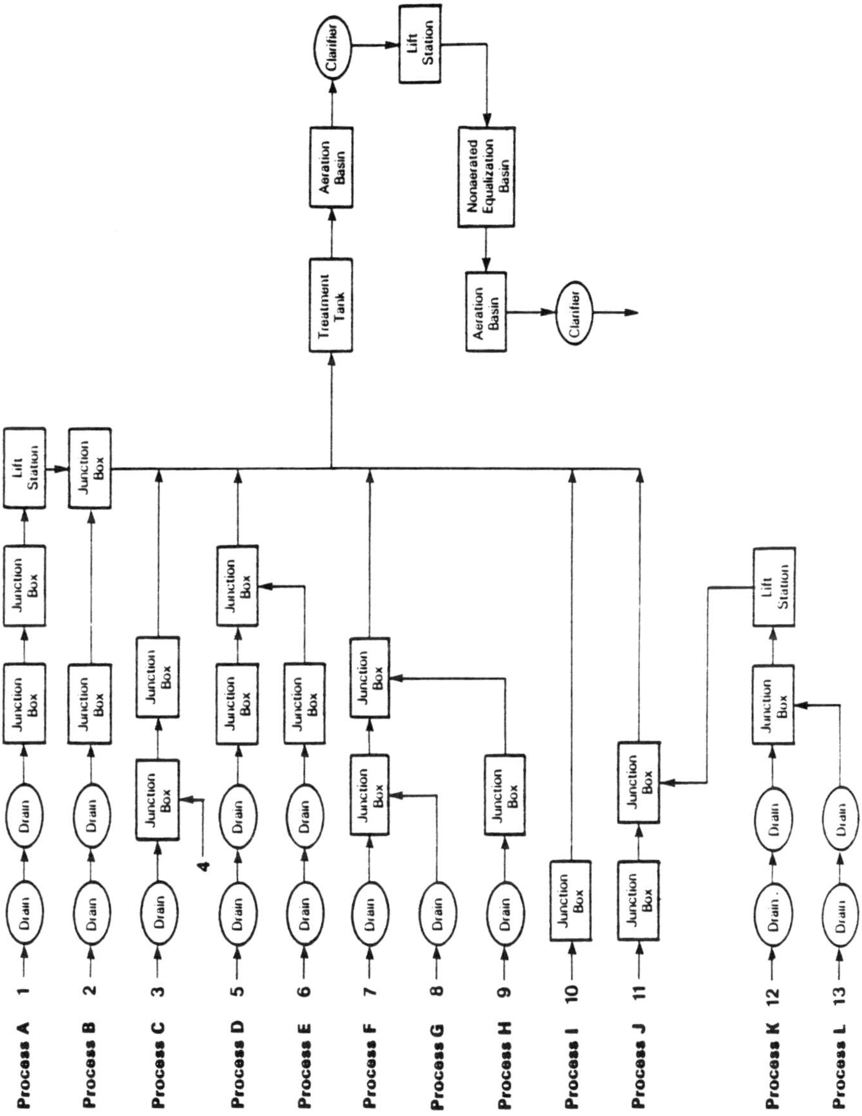

Figure F-3. Example Facility Three.
Treatment Configuration and Collection System.

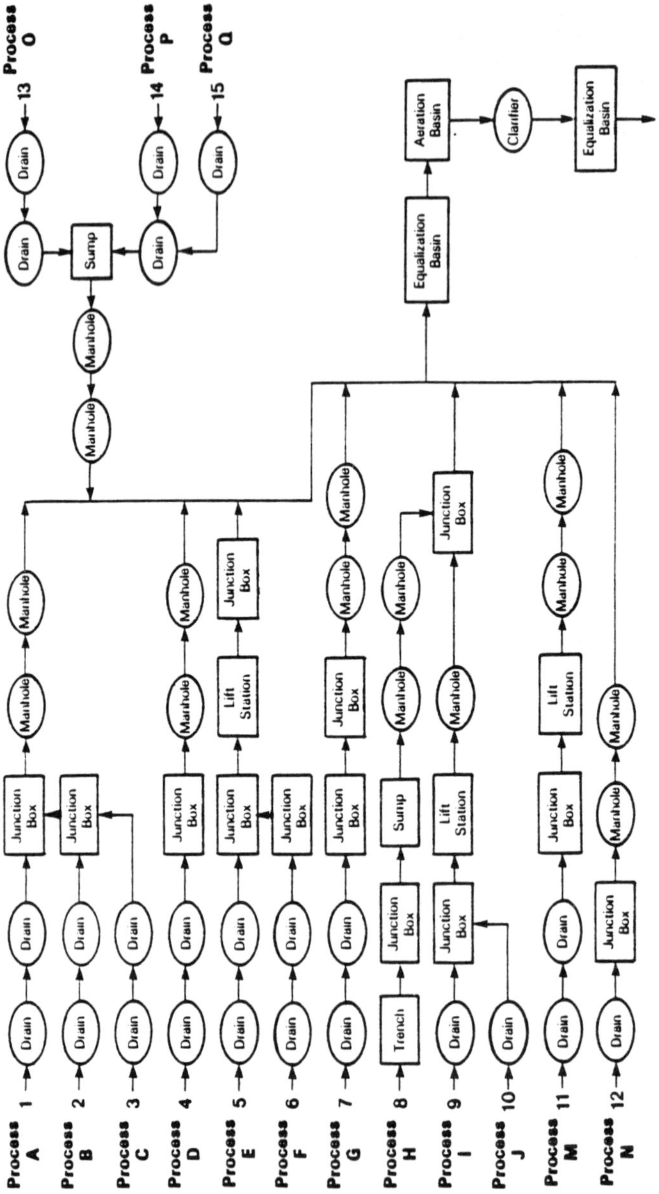

Figure F-4. Example Facility Four.
Treatment Configuration and Collection System.

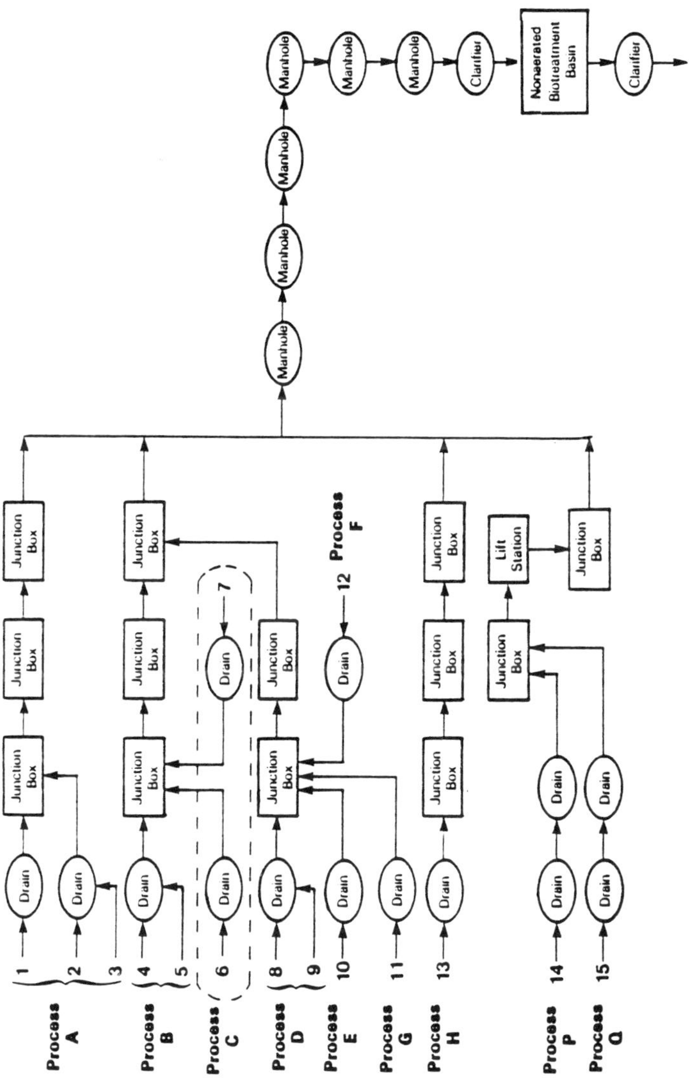

Figure F-5. Example Facility Five.
Treatment Configuration and Collection System.

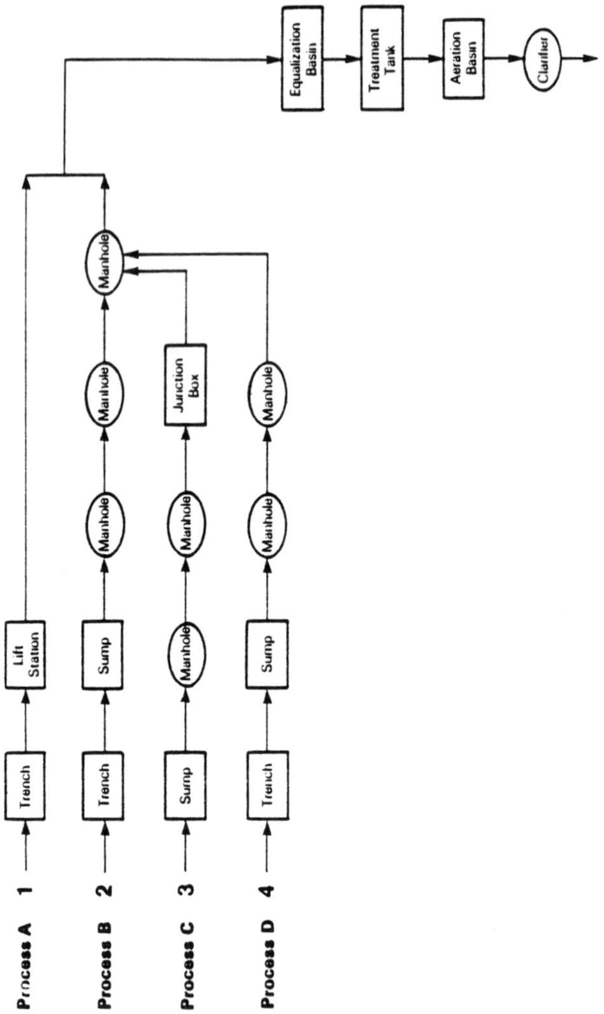

Figure F-6. Example Facility Six.
Treatment Configuration and Collection System.

F.3 EMISSIONS ESTIMATES

For each of the wastewater streams in the example facilities presented in the previous section, an estimate of the uncontrolled VOC emissions was developed. This section describes the development of these VOC emission estimates.

Estimated VOC emissions are a function of the configuration of the example facilities and of the properties of the specific organic compounds present in the wastewater streams. The fraction emitted (fe) for each compound within a wastewater stream is an estimate of the fraction of that compound that would be emitted from the wastewater stream during collection and treatment and is dependent on the Henry's Law Constant for that compound. In Chapter 3, fe was developed for five compounds that represent a range of volatility using several collection and treatment system configurations. The five compounds in order of increasing volatility were: 1) Phenol; 2) 1-Butanol; 3) Naphthalene; 4) Toluene; and 5) 1,3-Butadiene. Appendix A presents detailed calculations of fe for these five compounds for various collection and treatment configuration systems.

Correlations of fe as a function of Henry's Law constant were developed from the five compounds for each type of wastewater emission source. This was done for each collection and treatment source by constructing a plot of fe versus the Henry's Law constant for the five pollutants. These data were then fitted to a curve. The least-square curve fitting method was applied to all of the data and the best curve was chosen based on an optimum fit of $R^2 = 1.0$. (R^2 is a statistical measure of reliability of the relationship between the fraction emitted and Henry's Law constant values.) Table F-7 presents a summary of the results for these regressions, by collection and treatment source. (Weirs were not included in any of the example facility schematics.) Because these equations approximate fe, very high H values may result in fe values greater than 1. In addition, very low H values may result in fe values less than 0. When this occurs, fe is set to 1 for values greater than 1, and fe is set to 0 for values less than 0.

As a compound flows through the treatment scheme, a fraction of the compound is emitted into the surrounding air. The fraction of the compound entering the first source in the treatment scheme is 1.0. Some portion (fe_1) of this initial quantity is emitted from this first source in the

TABLE F-7. SUMMARY OF RESULTS-LEAST SQUARE FIT FOR fe VS. HENRY'S LAW CONSTANT

Source	Least Square Fit	R^2
Drain	fe = 0.0100 + 3.973 * (H)	0.997
Manhole	fe = 0.0003814 + 1.04 * (H)	1.000
Junction Box	fe = 1.658 * (H)0.592	0.902
Sump	fe = 0.007384 + 0.0004385 * ln (H)	0.876
Trench	fe = 0.06595 + 0.005006 * ln (H)	0.937
Lift Station	fe = 6.0118 * (H)0.6421	0.910
Oil-Water Separator	fe = 21.4034 * (H)0.8797	0.968
Equalization Basin - aerated	fe = 1.345 + 0.06927 * ln (H)	0.757
Equalization Basin - non-aerated	fe = 0.5752 + 0.0331 * ln (H)	0.844
Clarifier	fe = 0.03759 + 0.002351 * ln (H)	0.890
Treatment Tank	fe = 0.01374 + 0.0008558 ln (H)	0.889
Biobasin (aerated)	fe = 1.001 - 1.0028 * Fbioa	1.000
Biobasin (non-aerated)	fe = 0.7489 - 0.7883 * Fbion	0.940
Weir	fe = 422.8 * (Dw,i)0.6373	1.000

fe = fraction VOC emitted

H = Henry's Law Constant, atm . m^3/moL

R^2 = measure of the reliability of the relationship between the fe and H values. The optimum value for R^2 is 1.

collection/treatment scheme and the fraction pass through (fp_1=1-fe_1) enters the next source. This fraction pass through is multiplied by the fe from the second source (fe_2) to get the adjusted fraction emitted, Afe. The pass through of this second source is (fp_2) and is equal to (fp_1 - (fe_2 x fp_1)). Each source is treated in this manner until the compound reaches the point of discharge and becomes the effluent. The total or overall fraction emitted is the sum of Afe for each source.

There are two exceptions to this approach. If the source is an aerated or non-aerated biobasin, the biodegradation is assumed to be acting on each pollutant in the wastewater. If it is an oil-water separator, a fraction of each pollutant partitions in the oil layer on top of the wastewater and is removed in the recovered oil.

The degree of biodegradation for each pollutant was estimated using the draft CHEMDAT.7 program (developed by OAQPS[1]). The program was run for 8 different compounds representing 8 surrogate categories for both aerated and non-aerated biobasins. These categories were created for groupings of pollutant biorate and Henry's Law constant. The eight categories are: 1) High Henry's Law Constant, high biorate (HHHB); 2) High Henry's Law Constant, medium biorate (HHMB); 3) High Henry's Law Constant, low biorate (HHLB); 4) Medium Henry's Law Constant, high biorate (MHHB); 5) Medium Henry's Law Constant, medium biorate (MHMB); 6) Medium Henry's Law Constant, low biorate (MHLB); 7) Low Henry's Law constant, high biorate (LHHB); and 8) Low Henry's Law Constant, medium and low biorate (LHMB, LHLB). Table F-8 presents a summary of the surrogate categories and their corresponding fraction biodegraded. The ranges of Henry's Law Constant and biorate are also included for each surrogate category.

To compute the fraction pass-thru from an impoundment with biodegradation, an adjusted fraction emitted (Afe) and an adjusted fraction biodegraded (Afb) are calculated:

$$\text{Adjusted fe (Afe)} = fe_{component\ i} \times \text{pass-thru from previous source}$$
$$\text{Adjusted fb (Afb)} = fb_{component\ i} \times \text{pass-thru from previous source}$$

The fraction pass-thru from the biobasin is then calculated by subtracting the Afe and Afb from the influent pass-thru.

The amount of pollutant partitioning into the oil phase in an oil water separator was estimated using the octanol-water partition coefficient (K_{ow}).

TABLE F-8. SUMMARY OF SURROGATE CATEGORIES

Surrogate Category	Fraction Biodegradated (f_b) aerated	non-aerated
High Henry's Law Constant, high biorate (HHHB)	0.747	0.812
High Henry's Law Constant, medium biorate (HHMB)	0.814	0.519
High Henry's Law Constant, low biorate (HHLB)	0.638	0.071
Medium Henry's Law Constant, high biorate (MHHB)	0.966	0.824
Medium Henry's Law Constant, medium biorate (MHMB)	0.975	0.793
Medium Henry's Law Constant, low biorate (MHLB)	0.980	0.092
Low Henry's Law Constant, high biorate (LHHB)	0.998	0.950
Low Henry's Law Constant, medium and low biorate (LHMB, LHLB)	0.998	0.745

High Henry's Law Constant > 10^{-3}, atm m^3/moL

Medium Henry's Law Constant 10^{-3} to 10^{-5}, atm m^3/moL

Low Henry's Law Constant <10^{-5}, atm m^3/moL

High Biorate >10, mg vo/g biomass/hr

Medium Biorate 1 to 10 mg vo/g biomass/hr

Low Biorate <1, mg vo/g biomass/hr

K_{ow} is the solubility in oil divided by the solubility in water. This coefficient is estimated by the following equation:

$$K_{ow} = EXP\ (7.494 - \ln[Cs]) \qquad \text{Ref.3}$$

where: $$C_s = \text{solubility in water} \simeq \frac{\text{vapor pressure (mmHg)}}{760 * H\ (\text{atm m}^3/\text{moL})}$$

To determine the fraction removed in the oil-water separator the water in the oil-water separator was assumed to contain 1000 ppm oil. Therefore, the mass fraction of each organic compound in the oil is:

$$x_{oil} = 1 - [0.999/(0.999 + 0.001\ K_{ow})]$$

A more detailed description of this procedure is provided in "Distribution of VOC's in the Oil and Water Phases Using the Octanol-Water Coefficient", dated May 23, 1988.[2]

The fraction of VOC removed from the oil-water separator was assumed to be 80 percent of the VOC that partitioned into the oil phase.[3] The fraction VOC pass-thru from the oil water separator is the fraction pass-thru from the previous source, minus the adjusted fraction emitted from the oil water separator, minus the adjusted fraction VOC removed.

A sample calculation for Example Facility 1, Stream 1, is presented in Table F-9. (Figure F-1 presents the stream schematics for this facility.)

F.4 STEAM STRIPPER VOC REMOVAL

This section discusses the method used to predict VOC removal, and the associated emission reduction, through application of a steam stripper. The design of this steam stripper is presented in Chapter 4, and the associated control costs are shown in Chapter 6 of the attached document. Removals of individual compounds were predicted using Advanced Systems for Process Engineering (ASPEN), a computer software program.

Table F-10 presents the steam stripper removal efficiencies predicted by ASPEN for five compounds. From these data, an equation relating predicted removals to the compound's Henry's Law constant at 25°C was developed. This equation is also presented in Table F-10. Figure F-7 presents this relationship graphically.

TABLE F-9. UNCONTROLLED EMISSIONS - SAMPLE CALCULATION

EXAMPLE FACILITY: 1

STREAM: 1

FLOW RATE: 28.4 lpm

ORGANIC CONCENTRATION: 245 mg/l Ethylene oxide
2180 mg/l Acetaldehyde
9570 mg/l Ethylene glycol

Nomenclature

fe - mass fraction of VOC emitted

H - Henry's Law Constant, atm m^3/moL (T = 25°C)

fbion - mass fraction biodegraded in a non-aerated biobasin

fbioa - mass fraction biodegraded in an aerated biobasin

Afe - adjusted mass fraction of VOC emitted

fp - mass fraction of VOC pass through to next collection or treatment configuration

fb - adjusted mass fraction biodegraded in aerated or non-aerated biobasins

Cfe - cumulative mass fraction of VOC emitted

E - VOC emissions, Mg/yr

flow - flow rate of the stream, l/hr

conc - concentration of individual organic compound, mg/l

Properties	Ethylene oxide	Acetaldehyde	Ethylene glycol
H (atm-m^3/mol)	1.42E-04	9.50E-05	1.03E-07
fbioa	0.966	0.966	0.998

TABLE F-9. (Continued)

A. Fraction Emitted by Component	fe Ethylene oxide	fe Acetaldehyde	fe Ethylene glycol
Drain = 0.0100 + 3.973 * (H)	0.0106	0.0104	0.0100
Sump = 0.007384 + 0.0004385 * ln(H)	0.0350	0.0332	0.0329
Manhole = 0.0003814 + 1.04 * (H)	0.000529	0.000480	0.000382
Nonaerated Equalization Basin = 0.5752 + 0.0331 * ln(H)	0.282	0.269	0.0427
Aerated Biobasin = 1.001 - 1.0028 (fbioa)	0.0323	0.0323	0.000206
Clarifier = 0.03759 + 0.002351 * ln(H)	0.0168	0.0158	~0

B. Adjusted fraction emitted, (Afe)

$Afe_i = fp_{i-1} * fe_i$ $\qquad fp_o = 1.0$

$fp_i = f_{i-1} * Afe_i - fr - fb$

$fb_i = (fbion \text{ or } fbioa)(fp_i - 1)$

TABLE F-9. (Continued)

ETHYLENE OXIDE

Component	fe	Afe	fp	fb
Drain	0.0106	0.0106	0.989	---
Drain	0.0106	0.0105	0.979	---
Sump	0.0350	0.00343	0.975	---
Manhole	0.000529	0.000516	0.975	---
Manhole	0.000529	0.000515	0.974	---
Nonaerated Equalization Basin	0.282	0.275	0.700	---
Nonaerated Equalization Basin	0.282	0.197	0.502	---
Aerated Biobasin	0.0323	0.0162	7.85E-04	0.485
Clarifier	0.0168	1.32E-05	7.72E-04	---
Clarifier	0.0168	1.30E-05	7.59E-04	---

TABLE F-9. (Continued)

ACETALDEHYDE

Component	fe	Afe	fp	fb
Drain	0.0104	0.0104	0.990	---
Drain	0.0104	0.0103	0.979	---
Sump	0.00332	0.00325	0.976	---
Manhole	0.000480	0.000469	0.976	---
Manhole	0.000480	0.000468	0.975	---
Nonaerated Equalization Basin	0.269	0.262	0.713	---
Nonaerated Equalization Basin	0.269	0.262	0.521	---
Aerated Biobasin	0.0323	0.0168	9.14E-04	0.503
Clarifier	0.0158	1.44E-05	9.00E-04	---
Clarifier	0.0158	1.42E-05	8.85E-04	---

TABLE F-9. (Continued)

ETHYLENE GLYCOL

Component	fe	Afe	fp	fb
Drain	0.0100	0.0100	0.99	---
Drain	0.0100	0.00990	0.980	---
Sump	0.000329	0.000322	0.980	---
Manhole	0.000382	0.000374	0.979	---
Manhole	0.000382	0.000374	0.979	---
Nonaerated Equalization Basis	0.0427	0.0418	0.937	---
Nonaerated Equalization Basis	0.0427	0.0400	0.897	---
Aerated Biobasin	0.000206	0.000185	0.00182	0.895
Clarifier	~ 0	~ 0	0.00182	---
Clarifier	~ 0	~ 0	0.00182	---

TABLE F-9. (Continued)

D. Cumulative fraction emitted, CFE

$$Cfe = \sum_{i=1}^{n} Afe_i$$

n = number of VOC in the stream

	Cfe		
Component	Ethylene oxide	Acetaldehyde	Ethylene glycol
Drain	0.0106	0.0104	0.0100
Drain	0.0211	0.0207	0.0199
Sump	0.0245	0.0240	0.0202
Manhole	0.0250	0.0244	0.0206
Manhole	0.0256	0.0249	0.0210
Nonaerated Equalization Basin	0.301	0.287	0.0628
Nonaerated Equalization Basin	0.498	0.479	0.103
Aerated Biobasin	0.514	0.496	0.103
Clarifier	0.514	0.496	0.103
Clarifier	0.514	0.496	0.103

E (Uncontrolled Emissions in Mg/yr) = (fe * flow(l/hr)) * conc (mg/l) * 8760 yr/hr * 10^{-9} Mg/mg

$E_{Ethylene\ oxide}$ = (0.514 * 28.4 lpm * 60 min/hr) * (245 mg/l) * 8760 hr/yr * 10^{-9} Mg/mg = 1.88 Mg/yr

$E_{Acetaldehyde}$ = (0.496 * 28.4 lpm * 60 min/hr) * (2180 mg/l) * 8760 hr/yr * 10^{-9} Mg/mg = 16.1 Mg/yr

$E_{Ethylene\ glycol}$ = (0.103 * 28.4 lpm * 60 min/hr) * (9570 mg/l) * 8760 hr/yr * 10^{-9} Mg/mg = 14.7 Mg/yr

E_{VOC} = 32.7 Mg/yr

The equation predicts that compounds with a Henry's Law constant greater than or equal to 1.63×10^{-2} removal isessentially 100 percent. This equation was used to predict removals for all compounds in the example facility analysis.

The predicted removal efficiencies and uncontrolled emissions for the individual compounds were used to calculate the emission reduction and cost effectiveness associated with steam stripping each of the individual wastewater streams. A sample calculation of emission reduction is presented in Appendix A of the attached document.

F.5 TEST METHOD CONCENTRATION

In the example facility analysis, the VOC emissions (controlled and uncontrolled), the VOC emission reduction, and the cost effectiveness of control were calculated based on the actual organics present in the wastewater and the predicted fraction which would be emitted. However, the decision as to whether a stream should be controlled is based on the volatile organics measured using a test method. EPA has developed a draft test method to measure volatile organics known as Method 25D (see Appendix E). Test method procedures and operating parameters were chosen such that the test method detects the volatile organics that are expected to be emitted. The test method procedure calls for the analysis of the wastewater sample in a 50/50 by volume polyethylene glycol (PEG) and water dispersion medium. This sample is purged at a rate of 6 liters per minute for 30 minutes at 75 degrees Celsius.

The test method measures only carbons and halogens present in the wastewater and adjusts this based on an average molecular weight. Carbons are measured with a flame ionization detector (FID) and halogens (as chloride) are measured with a Hall electrolytic conductivity detector (HECD).

The Emissions Measurement Branch (EMB) has tested several compounds and percent recoveries, or the fraction measured by the method. Due to the limited number of pollutant results available, however, a theoretical estimation method was developed in order to estimate the percent recovery for the compounds in the example facility analysis. The theoretical method uses the compound structure to predict response correction factors for the FID and HECD and Henry's law constants in the PEG medium. From these estimates a percent recovery is predicted.

Table F-11 presents a summary of the predicted percent recoveries used for compounds contained in the example facility analysis.

TABLE F-10. STEAM STRIPPER REMOVAL EFFICIENCIES PREDICTED BY ASPEN

Compound	Henry's Law	Removal Efficiency %
1,3-Butadiene	0.142	100
Toluene	0.00668	100
Naphthalene	0.00118	99.9
1-Butanol	0.0000089	30.1
Phenol	0.000000454	2.2

f_r = 1.357 + 0.08677*ln (Henry's Law Constant, atm m^3/mol), R^2 = 0.887

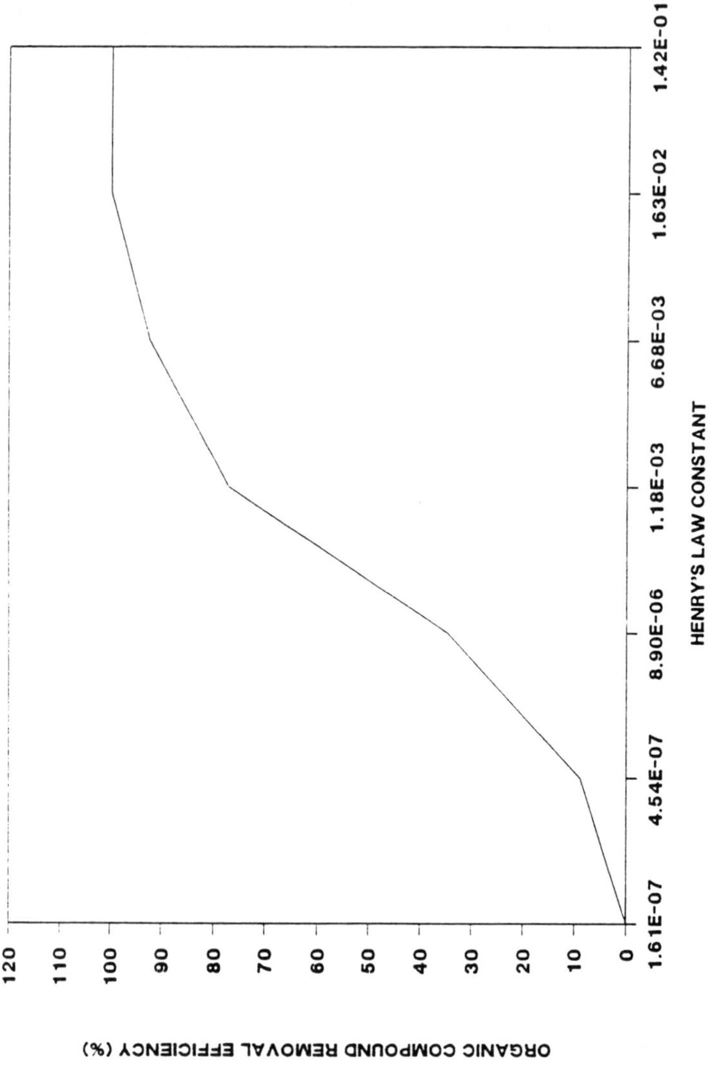

Figure F-7. Predicted steam stripper organic compound removal efficiencies based on Henry's Law constant for the compound.

TABLE F-11. PREDICTED TEST METHOD PERCENT RECOVERIES USED IN THE EXAMPLE FACILITIES ANALYSIS

Pollutant	Percent Recovery
1,3-Butadiene	119
1-Butanol	77
2,4-Dimethyl phenol	4.7
2,4-Dinitrophenol	1.4
Acenaphthene	123
Acetaldehyde	70
Acetone	89
Acetonitrile	74
Acrylonitrile	87
Aldicarb	2.2
Aniline	24.5
Benzene	122
Butyraldehyde	87
Carbon tetrachloride	103
Chloroaniline(p)	29
Chlorobenzene	116
Chloroethane	105
Chloroform	102.3
Chloronaphthalene(2)	120
Dibutyl phthalate	35.5
Dichlorobenzene(o)	113
Dichloroethane(1,2)	104
Dichloropropane(1,2)	106
Dichloropropylene	106
Diethylether	86
Dimethylamine	71
Di-n-octylphthalate	43
Ethylbenzene	120
Ethylene amines	71

TABLE F-11. (CONTINUED)

Pollutant	Percent Recovery
Ethylene dichloride	104
Ethylene glycol	0.42
Ethylene glycol monoethyl ether	23
Ethylene oxide	71
Fluoranthene	69
Hexane	108
Isobutyl alcohol	85
Isopropanol	79
Isopropyl acetate	79
Isopropylether	94
Methanol	32
Methyl chloride	104
Methylene chloride	102
Methyl isobutyl ketone	95
Naphthalene	124
n-Butyl acetate	81
n-Butyl alcohol	77
n-Propyl acetate	77
Phenol	5.7
Propionaldehyde	72
Propylene glycol	1.3
Propylene oxide	84
Pyridine	72
Styrene	123
Tetrachloroethane(1,1,2,2)	101
Tetrachloroethylene	105
Toluene	121
Trichloroethane(1,1,1)	82
Trichloroethane(1,1,2)	97
Trichloroethylene	105
Vinylidene chloride	106
Xylene	121

F.6 HAZARDOUS AIR POLLUTANTS

The primary focus in controlling VOC emissions is to reduce atmospheric ozone concentrations. However, other environmental benefits may result from the control of VOC. A large number of VOC are considered hazardous air pollutants. Of the 191 hazardous air pollutants listed in the proposed 1990 Clean Air Act (CAA) amendment, 90 percent are VOCs. Hazardous VOC are widely used in the industrial processes considered in this document. Seventy-nine percent of the VOC loadings in the example facility analysis are hazardous air pollutants. Therefore, reduction of VOC as illustrated with the addition of steam strippers to the example facilities results in a significant reduction of hazardous air pollutant emissions.

F.7 REFERENCES

1. Office of Air Quality Planning and Standards. U. S. Environmental Protection Agency. Hazardous Waste TSDF - Background Information for Proposed RCRA Air Emission standards. Volume 1 - Chapters 1 - 8, Appendices A-C. Review Draft. CHEMDAT.7 Computer Software Program. November 1988.

2. Memorandum to the file, "Distribution of VOC's in the Oil and Water Phases Using the Octanol-Water Coefficient," May 23, 1988.

3. Office of Air Quality Planning and Standards. VOC Emissions from Petroleum Refinery Wastewater Systems - Background Information for Proposed Standards. EPA-450/3-85-001a, February 1985, p. 3-43.

Glossary of Acronyms

API = American Petroleum Institute
BEC = Base Equipment Cost
CAA = Clean Air Act
CAS = Chemical Abstract Services
CO = Carbon monoxide
CTC = Control Technology Center
CWA = Clean Water Act
EPA = Environmental Protection Agency
ESP = Electrostatic Precipitator
fe = fraction emitted, emission factors
FR = Federal Register
H = Henry's Law Constant
H_2O = Water
lpm = liters per minute
MGD = Million Gallons per Day
NO_x = Nitrogen oxides
NPDES = National Pollutant Discharge Elimination System
NSPS = New Source Performance Standard
OAQPS = Office of Air Quality Planning and Standards
OCPSF = Organic Chemicals, Plastics, and Synthetic Fibers Manufacturing Industry
OSW = Office of Solid Waste
OWRS = Office of Water Regulation and Standards
PEC = Purchased Equipment Cost
PM = Particulate Matter
POTW = Publicly Owned Treatment Works
ppm = parts per million by weight
RACT = Reasonably Available Control Technology
SIC = Standard Industrial Classification
SO_2 = Sulfur dioxide
TAC = Total Annual Cost
TCI = Total Capital Investment

TSDF = Hazardous Waste Treatment, Storage, and Disposal Facilities

VO = Volatile Organics (this is the organic concentration as detected by draft EPA Reference Method 25D).

VOC = Volatile Organic Compounds (VOC) refers to all organic compounds except the following compounds that have been shown not to be photochemically reactive: methane, ethane, trichlorotrifluoroethane, methylene chloride, 1,1,1-trichloroethane, trichlorofluoromethane, dichlorodifluoromethane, chlorodifluoromethane, trifluoromethane, dichlorotetrafluoroethane, and chloropentafluoroethane. (See 45 FR 48941, July 22, 1980.)

Other Noyes Publications

CONTAMINATION OF GROUND WATER
Prevention, Assessment, Restoration

by

Michael Barcelona, Allen Wehrmann
Illinois State Water Survey

Joseph F. Keely
U.S.E.P.A.

Wayne A. Pettyjohn
Oklahoma State University

Pollution Technology Review No. 184

The need exists for a resource document that brings together available technical information on ground-water management in a form convenient for ground-water personnel at all levels of government, as well as in the private sector. The information contained in this handbook of ground-water contamination meets that need. The book covers measures for (1) prevention of contamination, (2) assessment of extent of contamination, and (3) restoration of ground-water quality.

Because contamination of ground water has occurred in every state and is being detected with increasing frequency, regulatory agencies and courts have been developing guidelines, laws and rules to protect this resource. Ground-water quality laws deal with both the prevention of ground-water contamination and assigning responsibility for ground-water protection or cleanup and legal liability for damages where contamination has occurred. Provisions aimed at prevention of contamination regulate the conduct of activities which could have the effect of polluting ground-water or posing risks to human health. Other statutory provisions call for government or private party response to incidents of contamination; they also may assign penalties or other legal liability to polluters. The operation of these provisions is generally triggered by the release of certain harmful substances, identified by statute or government regulation, into the environment.

This book will provide research information to decision makers, field managers, and the scientific community which will help satisfy the immediate need for technology transfer applicable to ground-water contamination control and prevention.

A condensed contents listing **part titles, chapter titles, and selected subtitles** is given below.

ISBN 0-8155-1243-0 (1990) **7" x 10"** 213 pages

Other Noyes Publications

HOW TO MEET REQUIREMENTS FOR HAZARDOUS WASTE LANDFILL DESIGN, CONSTRUCTION AND CLOSURE

U.S. Environmental Protection Agency

Pollution Technology Review No. 185

The U.S. Environmental Protection Agency's (EPA's) minimum technological requirements for hazardous waste landfill design were set forth by Congress in the 1984 Hazardous and Solid Waste Amendments (HSWA). HSWA covered requirements for landfill liners and leachate collection and removal systems, as well as leak detection systems for landfills, surface impoundments, and waste piles. In response to HSWA and other Congressional mandates, EPA has issued proposed regulations and guidance on the design of these systems, and on construction, quality assurance, final cover and response action plans for responding to landfill leaks.

This book outlines in detail the provisions of the minimum technology guidance and proposed regulations, and **offers practical and detailed technology transfer information** on the construction of hazardous waste facilities that comply with these requirements.

A condensed table of contents listing **chapter titles and selected subtitles** is given below.

ISBN 0-8155-1242-2 (1990) **7" x 10"** 123 pages

OSHA REGULATED HAZARDOUS SUBSTANCES

Health, Toxicity, Economic and Technological Data

Occupational Safety and Health Administration
U.S. Department of Labor

This two-volume book provides industrial exposure data and control technologies for more than 650 substances currently regulated, or candidates for regulation, by the Occupational Safety and Health Administration (OSHA). The health, toxicity, economic and technological data provided for each substance are intended to serve as a reference for those who are potentially exposed to one or more of these substances in their workplace, or for those who have supervisory or management responsibility for workers potentially exposed. OSHA "permissible exposure limits" (PELs) for these 650 substances reflect all updates and changes as presented in the *Federal Register.*

The information on each substance in the book includes, as available, synonyms, trade names, physical description, health effects, toxicity/exposure limits, industry use data, and NIOSH National Occupational Exposure Survey data, NIOSH National Occupational Hazard Survey data, OSHA/exposure data, engineering controls, personal protective equipment, and storage.

Indexes included in the book provide cross referencing by synonyms, trade name, and Chemical Abstract Service (CAS) number.

Below is an alphabetical listing of the first 61 substances in the book.

Acetaldehyde
Acetic acid
Acetic anhydride
Acetone
Acetonitrile
2-Acetylaminofluorine
Acetylene
Acetylene tetrabromide
Acetylsalicylic acid
Acrolein
Acrylamide
Acrylic acid
Acrylonitrile
Aldrin
Allyl alcohol
Allyl chloride
Allyl glycidyl ether
Allyl propyl disulfide
alpha-Alumina
Aluminum: alkyls
Aluminum: metal and oxide
Aluminum: pyro powders
Aluminum: soluble salts
Aluminum: welding fumes
4-Aminodiphenyl
2-Aminopyridine
Amitrole
Ammonia
Ammonium chloride (fume)
Ammonium perfluorooctanoate
Ammonium sulfamate
Amosite
n-Amyl acetate
sec-Amyl acetate
Aniline and homologues
Anisidine (o-, p- isomers)
Antimony and compounds
Antimony trioxide as Sb
ANTU
Argon
Arsenic and compounds, as As
Arsenic trioxide production
Arsine
Asbestos
Asphalt fumes
Atrazine
Azinphos-methyl
Barium sulfate
Barium, soluble compounds
Benomyl
Benzene
Benzidine
Benzidine (based dyes)
Benzo(a)pyrene
Benzoyl peroxide
Benzyl chloride
Beryllium and compounds
Biphenyl
Bismuth telluride
Borates, tetra, sodium salts
Boron oxide
plus 589 other substances

ISBN 0-8155-1240-6 (1990) **6" x 9"** **2 volumes** 2294 pages